Texts in Theoretical Computer Science
An EATCS Series

For further volumes:
http://www.springer.com/series/3214

Fedor V. Fomin · Dieter Kratsch

Exact Exponential
Algorithms

 Springer

Prof. Dr. Fedor V. Fomin
University of Bergen
Inst. Informatics
PO Box 7800
5020 Bergen
Norway
fomin@ii.uib.no

Prof. Dieter Kratsch
Université Paul Verlaine - Metz
LITA, UFR MIM
Dépt. Informatique
Ile du Saulcy
57045 Cedex 1
France
kratsch@univ-metz.fr

Series Editors
Prof. Dr. Wilfried Brauer
Institut für Informatik der TUM
Boltzmannstr. 3
85748 Garching, Germany
brauer@informatik.tu-muenchen.de

Prof. Dr. Juraj Hromkovič
ETH Zentrum
Department of Computer Science
Swiss Federal Institute of Technology
8092 Zürich, Switzerland
juraj.hromkovic@inf.ethz.ch

Prof. Dr. Grzegorz Rozenberg
Leiden Institute of Advanced
Computer Science
University of Leiden
Niels Bohrweg 1
2333 CA Leiden, The Netherlands
rozenber@liacs.nl

Prof. Dr. Arto Salomaa
Turku Centre of Computer Science
Lemminkäisenkatu 14 A
20520 Turku, Finland
asalomaa@utu.fi

ISSN 1862-4499
ISBN 978-3-642-16532-0   e-ISBN 978-3-642-16533-7
DOI 10.1007/978-3-642-16533-7
Springer Heidelberg Dordrecht London New York

ACM Codes: F.2, G.1, G.2

Cover design: KuenkelLopka GmbH, Heidelberg

Printed on acid-free paper

Springer is part of Springer Science+Business Media (www.springer.com)

Preface

For a long time computer scientists have distinguished between fast and slow algorithms. Fast (or good) algorithms are the algorithms that run in polynomial time, which means that the number of steps required for the algorithm to solve a problem is bounded by some polynomial in the length of the input. All other algorithms are slow (or bad). The running time of slow algorithms is usually exponential. *This book is about bad algorithms.*

There are several reasons why we are interested in exponential time algorithms. Most of us believe that there are many natural problems which cannot be solved by polynomial time algorithms. The most famous and oldest family of hard problems is the family of NP-complete problems. Most likely there are no polynomial time algorithms solving these hard problems and in the worst-case scenario the exponential running time is unavoidable.

Every combinatorial problem is solvable in finite time by enumerating all possible solutions, i.e. by brute-force search. But is brute-force search always unavoidable? Definitely not. Already in the nineteen sixties and seventies it was known that some NP-complete problems can be solved significantly faster than by brute-force search. Three classic examples are the following algorithms for the TRAVELLING SALESMAN problem, MAXIMUM INDEPENDENT SET, and COLORING. The algorithm of Bellman [17] and Held and Karp [111] from 1962 solves the TRAVELLING SALESMAN problem with n cities in time $\mathcal{O}(2^n n^2)$, which is much faster than the trivial $\mathcal{O}(n!n)$ brute-force search. In 1977, Tarjan and Trojanowski [213] gave an $\mathcal{O}(2^{n/3})$ time algorithm computing a maximum independent set in a graph on n vertices, improving on the brute-force search algorithm which takes $\mathcal{O}(n2^n)$. In 1976, Lawler [150] constructed an $\mathcal{O}(n(1 + \sqrt[3]{3})^n)$ time algorithm solving the COLORING problem for which brute-force solution runs in time $\mathcal{O}(n^{n+1})$. On the other hand, for some NP-complete problems, like SATISFIABILITY, regardless of all developments in algorithmic techniques in the last 50 years, we know of no better algorithm than the trivial brute-force search that tries all possible solutions. It is a great intellectual challenge to find whether enumeration of solutions is the only approach to solve NP problems in general.[1]

With the development of the area of exact algorithms, it has become possible to improve significantly the running time of the classical algorithms for MAXIMUM INDEPENDENT SET and GRAPH COLORING. Moreover, very often the techniques for solving NP problems can be used to solve problems that are presumably harder than NP-complete problems, like #P and PSPACE-complete problems. On the other hand, for some problems, like the TRAVELLING SALESMAN problem, no progress has been made in the last 50 years. To find an explanation of the wide variation in the worst-case complexities of known exact algorithms is another challenge.

Intellectual curiosity is not the only reason for the study of exponential algorithms. There are certain applications that require exact solutions of NP-hard problems, although this might only be possible for moderate input sizes. And in some situations the preference of polynomial time over exponential is debatable. Richard Lipton in his blog *"Gödel's Lost Letter and P$\neq$NP"*[2], attributes the following saying to Alan Perlis, the first Turing Award winner, *"for every polynomial-time algorithm you have, there is an exponential algorithm that I would rather run"*. The point is simple: $n^3 > 1.0941^n \cdot n$ for $n \leq 100$ and on instances of moderate sizes, the exponential time algorithm can be preferable to polynomial time algorithms. And a reduction of the base of the exponential running time, say from $\mathcal{O}(1.8^n)$ to $\mathcal{O}(1.7^n)$, increases the size of the instances solvable within a given amount of time by a constant *multiplicative* factor; running a given exponential algorithm on a faster computer can enlarge the size only by a (small) *additive* factor. *Thus "bad" exponential algorithms are not that bad in many situations!*

And the final reason of our interest in exact algorithms is of course that the design and analysis of exact algorithms leads to a better understanding of NP-hard problems and initiates interesting new combinatorial and algorithmic challenges.

Our interest in exact algorithms was attracted by an amazing survey by Gerhard Woeginger [220]. This influential survey fascinated many researchers and, we think, was one of the main reasons why the area of exact algorithms changed drastically in the last decade. Still being in a nascent stage, the study of exact algorithms is now a dynamic and vibrant area. While there are many open problems, and new techniques to solve these problems are still appearing, we believe it is the right moment to summarize the work on exact algorithms in a book. The main intention of this book is to provide an introduction to the area and explain the most common algorithmic

---

[1] The origin of this question can be traced to the famous letter of Gödel to von Neumann from 1956:

> "Es wäre interessant zu wissen, ... wie stark im allgemeinen bei finiten kombinatorischen Problemen die Anzahl der Schritte gegenüber dem blossen Probieren verringert werden kann."

English translation: It would be interesting to know, ... how strongly in general the number of steps in finite combinatorial problems can be reduced with respect to simple exhaustive search. [205]

[2] Weblog post *Fast Exponential Algorithms* from February 13, 2009, available at http://rjlipton.wordpress.com/

techniques. We have tried to make the results accessible not only to the specialists working in the area but to a more general audience of students and researchers in Computer Science, Operations Research, Optimization, Combinatorics, and other fields related to algorithmic solutions of hard optimization problems. Therefore, our preferences in presenting the material were for giving the basic ideas, avoiding improvements which require significant technical effort.

Bergen, Metz,   *Fedor V. Fomin*
August 2010   *Dieter Kratsch*

Acknowledgements

We are deeply indebted to Gerhard Woeginger for attracting our attention to Exact Algorithms in 2002. His survey on exact algorithms for NP-hard problems was a source of inspiration and triggered our interest in this area.

This work would be impossible without the help and support of many people. We are grateful to all our coauthors in the area of exact algorithms: Omid Amini, Hans Bodlaender, Hajo Broersma, Jianer Chen, Frederic Dorn, Mike Fellows, Henning Fernau, Pierre Fraigniaud, Serge Gaspers, Petr Golovach, Fabrizio Grandoni, Gregory Gutin, Frédéric Havet, Pinar Heggernes, Pim van 't Hof, Kjartan Høie, Iyad A. Kanj, Joachim Kneis, Arie M. C. A. Koster, Jan Kratochvíl, Alexander Langer, Mathieu Liedloff, Daniel Lokshtanov, Frédéric Mazoit, Nicolas Nisse, Christophe Paul, Daniel Paulusma, Eelko Penninkx, Artem V. Pyatkin, Daniel Raible, Venkatesh Raman, Igor Razgon, Frances A. Rosamond, Peter Rossmanith, Saket Saurabh, Alexey A. Stepanov, Jan Arne Telle, Dimitrios Thilikos, Ioan Todinca, Yngve Villanger, and Gerhard J. Woeginger. It was a great pleasure to work with all of you and we profited greatly from all this cooperation. Special thanks go to Fabrizio Grandoni for introducing us to the world of Measure & Conquer.

Many of our colleagues helped with valuable comments, corrections and suggestions. We are grateful for feedback from Evgeny Dantsin, Serge Gaspers, Fabrizio Grandoni, Petteri Kaski, Mikko Koivisto, Alexander Kulikov, Mathieu Liedloff, Daniel Lokshtanov, Jesper Nederlof, Igor Razgon, Saket Saurabh, Uwe Schöning, Gregory Sorkin, Ioan Todinca, K. Venkata, Yngve Villanger, Ryan Williams, and Peter Rossmanith.

It is a special pleasure to thank our wives Nora and Tanya. Without their encouragement this work would have taken much longer.

Contents

Chapter 1
Introduction

In this introductory chapter we start with a preliminary part and present then two classical exact algorithms breaking the triviality barrier. The first one, from the nineteen sixties, is the dynamic programming algorithm of Bellman, Held and Karp to solve the TRAVELLING SALESMAN problem [16, 17, 111]. The second is a branching algorithm to compute a maximum independent set of a graph. The main idea of this algorithm can be traced back to the work of Miller and Muller [155] and Moon and Moser [161] from the nineteen sixties.

The history of research on exact algorithms for these two NP-hard problems is contrasting. Starting with the algorithm of Tarjan and Trojanowski [213] from 1977 there was a chain of dramatic improvements in terms of the running time of an algorithm for the MAXIMUM INDEPENDENT SET problem. For the TRAVELLING SALESMAN problem, despite many attempts, no improvement on the running time of the Bellman-Held-Karp's algorithm was achieved so far.

1.1 Preliminaries

$\mathcal{O}^*$ *notation.* The classical big-O notation is defined as follows. For functions $f(n)$ and $g(n)$ we write $f = \mathcal{O}(g)$ if there are positive numbers n_0 and c such that for every $n > n_0$, $f(n) < c \cdot g(n)$. In this book we use a modified big-O notation that suppresses all polynomially bounded factors. For functions f and g we write $f(n) = \mathcal{O}^*(g(n))$ if $f(n) = \mathcal{O}(g(n)poly(n))$, where $poly(n)$ is a polynomial. For example, for $f(n) = 2^n n^2$ and $g(n) = 2^n$, $f(n) = \mathcal{O}^*(g(n))$. This modification of the classical big-O notation can be justified by the exponential growth of $f(n)$. For instance, the running time $(\sqrt{2})^n poly(n)$ is sandwiched between running times 1.4142135^n and 1.4142136^n for every polynomial $poly(n)$ and sufficiently large n. In many chapters of this book when estimating the running time of algorithms, we have exponential functions where the base of the exponent is some real number. Very often we round the base of the exponent up to the fourth digit after the decimal point. For example, for running time $\mathcal{O}((\sqrt{2})^n)$, we have $\sqrt{2} = 1.414213562...$, and $(\sqrt{2})^n poly(n) =$

F.V. Fomin, D. Kratsch, *Exact Exponential Algorithms*, Texts in Theoretical
Computer Science. An EATCS Series, DOI 10.1007/978-3-642-16533-7_1,
© Springer-Verlag Berlin Heidelberg 2010

$\mathcal{O}(1.4143^n)$. Hence when we round reals in the base of the exponent, we use the classical big-O notation. We also write $f = \Omega(g)$, which means that $g = \mathcal{O}(f)$, and $f = \Theta(g)$, which means that $f = \Omega(g)$ and $f = \mathcal{O}(g)$.

Measuring quality of exact algorithms. The common agreement in polynomial time algorithms is that the running time of an algorithm is estimated by a function either of the input length or of the input "size". The input length can be defined as the number of bits in any "reasonable" encoding of the input over a finite alphabet; but the notion of input size is problem dependent. Usually every time we speak about the input size, we have to specify what we mean by that. Let us emphasize that for most natural problems the length of the input is not exactly the same as what we mean by its "size". For example, for a graph G on n vertices and m edges, we usually think of the size of G as $\Theta(n+m)$, while the length (or the number of bits) in any reasonable encoding over a finite alphabet is $\Theta(n+m\log n)$. Similarly for a CNF Boolean formula F with n variables and m clauses, the size of F is $\Theta(n+m)$ and the input length is $\Theta(n+m\log m)$.

So what is the appropriate input "size" for exponential time algorithms? For example for an optimization problem on graphs, the input "size" can be the number of vertices, the number of edges or the input length. In most parts of this book we follow the more or less established tradition that

- Optimization problems on graphs are analyzed in terms of the number of vertices;
- Problems on sets are analyzed in terms of the number of elements;
- Problems on Boolean formulas are analyzed in terms of the number of variables.

An argument for such choices of the "size" is that with such parameterization it is often possible to measure the improvement over the trivial brute-force search algorithm. Every search version of the problem L in NP can be formulated in the following form:

Given x, find y so that $|y| \leq m(x)$ and $R(x,y)$ (if such y exists).

Here x is an instance of L, $|y|$ is the length (the number of bits in the binary representation) of certificate y, $R(x,y)$ is a polynomial time decidable relation that verifies the certificate y for instance x, and $m(x)$ is a polynomial time computable and polynomially bounded complexity parameter that bounds the length of the certificate y. Thus problem L can be solved by enumerating all possible certificates y of length at most $m(x)$ and checking for each certificate in polynomial time if $R(x,y)$. Therefore, the running time of the brute-force search algorithm is up to a polynomial multiplicative factor proportional to the number of all possible certificates of length at most $m(x)$, which is $\mathcal{O}^*(2^{m(x)})$.

Let us give some examples.

- *Subset problems.* In a subset problem every feasible solution can be specified as a subset of an underlying ground set. If the cardinality of the ground set is n, then every subset S of the ground set can be encoded by a binary string of length n. The ith element of the string is 1 if and only if the ith element of the instance x is in S. In this case $m(x) = n$ and the brute-force search can be done in time $\mathcal{O}^*(2^n)$. For instance, a truth assignment in the SATISFIABILITY problem corresponds to selecting a subset of TRUE variables. A candidate solution in this case is the subset of variables, and the size of each subset does not exceed the number of variables, hence the length of the certificate does not exceed n. Thus the brute-force search enumerating all possible subsets of variables and checking (in polynomial time) whether the selected assignment satisfies the formula takes $\mathcal{O}^*(2^n)$ steps. In the MAXIMUM INDEPENDENT SET problem, every subset of the vertex set is a solution candidate of size at most n, where n is the number of vertices of the graph. Again, the brute-force search for MAXIMUM INDEPENDENT SET takes $\mathcal{O}^*(2^n)$ steps.

- *Permutation problems.* In a permutation problem every feasible solution can be specified as a total ordering of an underlying ground set. For instance, in the TRAVELLING SALESMAN problem, every tour corresponds to a permutation of the cities. For an instance of the problem with n cities, possible candidate solutions are ordered sets of n cities. The size of the candidate solution is n and the number of different ordered sets of size n is $n!$. In this case $m(x) = \log_2 n!$ and the trivial algorithm runs in time $\mathcal{O}^*(n!)$.

- *Partition problems.* In a partition problem, every feasible solution can be specified as a partition of an underlying ground set. An example of such a problem is the GRAPH COLORING problem, where the goal is to partition the vertex set of an n-vertex graph into color classes. In this case $m(x) = \log_2 n^n$ and the brute-force algorithm runs in $\mathcal{O}^*(n^n) = \mathcal{O}^*(2^{n \log n})$ time.

Intuitively, such a classification of the problems according to the number of candidate solutions creates a complexity hierarchy of problems, where subset problems are "easier" than permutation problems, and permutation problems are "easier" than partition problems. However, we do not have any evidences that such a hierarchy exists; moreover there are permutation problems solvable in time $\mathcal{O}^*((2-\varepsilon)^n)$ for some $\varepsilon > 0$. There are also some subset problems for which we do not know anything better than brute-force search. We also should say that sometimes such classification is ambiguous. For example, is the HAMILTONIAN CYCLE problem a permutation problem for vertices or a subset problem for edges? One can argue that on graphs, where the number of edges m is less than $\log_2 n!$, the algorithm trying all possible edge subsets in time $\mathcal{O}^*(2^m)$ is faster than $\mathcal{O}^*(n!)$, and in these cases we have to specify what we mean by the brute-force algorithm. Fortunately, such ambiguities do not occur often.

Parameterized complexity. The area of exact exponential algorithms is not the only one dealing with exact solutions of hard problems. The parameterized complexity theory introduced by Downey and Fellows [66] is a general framework for a refined analysis of hard algorithmic problems. Parameterized complexity measures complexity not only in terms of input length but also in terms of a parameter which is a numerical value not necessarily dependent on the input length. Many parameterized algorithmic techniques evolved accompanied by a powerful complexity theory. We refer to recent monographs of Flum and Grohe [78] and Niedermeier [164] for overviews of parameterized complexity. Roughly speaking, parameterized complexity seeks the possibility of obtaining algorithms whose running time can be bounded by a polynomial function of the input length and, usually, an exponential function of the parameter. Thus most of the exact exponential algorithms studied in this book can be treated as parameterized algorithms, where the parameter can be the number of vertices in a graph, the number of variables in a formula, etc. However, such a parameterization does not make much sense from the point of view of parameterized complexity, where the fundamental assumption is that the parameter is independent of the input size. In particular, it is unclear whether the powerful tools from parameterized complexity can be used in this case. On the other hand, there are many similarities between the two areas, in particular some of the basic techniques like branching, dynamic programming, iterative compression and inclusion-exclusion are used in both areas. There are also very nice connections between subexponential complexity and parameterized complexity.

1.2 Dynamic Programming for TSP

Travelling Salesman Problem. In the TRAVELLING SALESMAN problem (TSP), we are given a set of distinct cities $\{c_1, c_2, \ldots, c_n\}$ and for each pair $c_i \neq c_j$ the distance between c_i and c_j, denoted by $d(c_i, c_j)$. The task is to construct a tour of the travelling salesman of minimum total length which visits all the cities and returns to the starting point. In other words, the task is to find a permutation π of $\{1, 2, \ldots, n\}$, such that the following sum is minimized

$$\sum_{i=1}^{n-1} d(c_{\pi(i)}, c_{\pi(i+1)}) + d(c_{\pi(n)}, c_{\pi(1)}).$$

How to find a tour of minimum length? The easy way is to generate all possible solutions. This requires us to verify all permutations of the cities and the number of all permutations is $n!$. Thus a naive approach here requires at least $n!$ steps. Using dynamic programming one obtains a much faster algorithm.

The dynamic programming algorithm for TSP computes for every pair (S, c_i), where S is a nonempty subset of $\{c_2, c_3, \ldots, c_n\}$ and $c_i \in S$, the value $OPT[S, c_i]$ which is the minimum length of a tour which starts in c_1, visits all cities from S and ends in c_i. We compute the values $OPT[S, c_i]$ in order of increasing cardinality of S. The computation of $OPT[S, c_i]$ in the case S contains only one city is trivial, because in this case, $OPT[S, c_i] = d(c_1, c_i)$. For the case $|S| > 1$, the value of $OPT[S, c_i]$ can be expressed in terms of subsets of S:

$$OPT[S, c_i] = \min\{OPT[S \setminus \{c_i\}, c_j] + d(c_j, c_i) : c_j \in S \setminus \{c_i\}\}. \qquad (1.1)$$

Indeed, if in some optimal tour in S terminating in c_i, the city c_j immediately precedes c_i, then

$$OPT[S,c_i] = OPT[S \setminus \{c_i\}, c_j] + d(c_j, c_i).$$

Thus taking the minimum over all cities that can precede c_i, we obtain (1.1). Finally, the value OPT of the optimal solution is the minimum of

$$OPT[\{c_2, c_3, \ldots, c_n\}, c_i] + d(c_i, c_1),$$

where the minimum is taken over all indices $i \in \{2, 3, \ldots, n\}$.

Such a recurrence can be transformed in a dynamic programming algorithm by solving subproblems in increasing sizes, which here is the number of cities in S. The corresponding algorithm tsp is given in Fig. 1.1.

Algorithm tsp($\{c_1, c_2, \ldots c_n\}, d$).
Input: Set of cities $\{c_1, c_2, \ldots, c_n\}$ and for each pair of cities c_i, c_j the distance $d(c_i, c_j)$.
Output: The minimum length of a tour.

> **for** $i = 2$ *to* n **do**
> $\quad \lfloor \; OPT[c_i, c_i] = d(c_1, c_i)$
>
> **for** $j = 2$ *to* $n - 1$ **do**
> $\quad$ **forall** $S \subseteq \{2, 3, \ldots, n\}$ *with* $|S| = j$ **do**
> $\quad \quad \lfloor \; OPT[S, c_i] = \min\{OPT[S \setminus \{c_i\}, c_k] + d(c_k, c_i) : c_k \in S \setminus \{c_i\}\}$
>
> **return** $\min\{OPT[\{c_2, c_3, \ldots, c_n\}, c_i] + d(c_i, c_1) : i \in \{2, 3, \ldots, n\}\}$

Fig. 1.1 Algorithm tsp for the TRAVELLING SALESMAN problem

Before analyzing the running time of the dynamic programming algorithm let us give a word of caution. Very often in the literature the running time of algorithms is expressed in terms of basic computer primitives like arithmetic (add, subtract, multiply, comparing, floor, etc.), data movement (load, store, copy, etc.), and control (branching, subroutine call, etc.) operations. For example, in the unit-cost random-access machine (RAM) model of computation, each of such steps takes constant time. The unit-cost RAM model is the most common model appearing in the literature on algorithms. In this book we also adapt the unit-cost RAM model and treat these primitive operations as single computer steps. However in some parts of the book dealing with computations with huge numbers such simplifying assumptions would be too inaccurate.

The reason is that in all known realistic computational models arithmetic operations with two b-bit numbers require time $\Omega(b)$, which brings us to the log-cost RAM model. For even more realistic models one has to assume that two b-bit integers can be added, subtracted, and compared in $\mathcal{O}(b)$ time, and multiplied in $\mathcal{O}(b \log b \log \log b)$ time. But this level of precision is not required for most of the results discussed in this book. Because of the $\mathcal{O}^*$-notation, we can neglect the difference between log-cost and unit-cost RAM for most of the algorithms presented in this book. Therefore, normally we do not mention the model used to analyze running times of algorithms (assuming unit-cost RAM model), and specify it only when the difference between computational models becomes important.

Let us come back to TSP. The amount of steps required to compute (1.1) for a fixed set S of size k and all vertices $c_i \in S$ is $\mathcal{O}(k^2)$. The algorithm computes (1.1) for every subset S of cities, and thus takes time $\sum_{k=1}^{n-1} \mathcal{O}(\binom{n}{k})$. Therefore, the total time to compute OPT is

$$\sum_{k=1}^{n-1} \mathcal{O}(\binom{n}{k} k^2) = \mathcal{O}(n^2 2^n).$$

The improvement from $\mathcal{O}(n!n)$ in the trivial enumeration algorithm to $\mathcal{O}^*(2^n)$ in the dynamic programming algorithm is quite significant.

For the analyses of the TSP algorithm it is also important to specify which model is used. Let W be the maximum distance between the cities. The running time of the algorithm for the unit-cost RAM model is $\mathcal{O}^*(2^n)$. However, during the algorithm we have to operate with $\mathcal{O}(\log nW)$-bit numbers. By making use of more accurate log-cost RAM model, we estimate the running time of the algorithm as $2^n \log W n^{\mathcal{O}(1)}$. Since W can be arbitrarily large, $2^n \log W n^{\mathcal{O}(1)}$ is not in $\mathcal{O}^*(2^n)$.

Finally, once all values $OPT[S, c_i]$ have been computed, we can also construct an optimal tour (or a permutation π) by making use of the following observation: A permutation π, with $\pi(c_1) = c_1$, is optimal if and only if

$$OPT = OPT[\{c_{\pi(2)}, c_{\pi(3)}, \ldots, c_{\pi(n)}\}, c_{\pi(n)}] + d(c_{\pi(n)}, c_1),$$

and for $k \in \{2, 3, \ldots, n-1\}$,

$$OPT[\{c_{\pi(2)}, \ldots, c_{\pi(k+1)}\}, c_{\pi(k+1)}] = OPT[\{c_{\pi(2)}, \ldots, c_{\pi(k)}\}, c_{\pi(k)}]$$
$$+ d(c_{\pi(k)}, c_{\pi(k+1)}).$$

A dynamic programming algorithm computing the optimal value of the solution of a problem can typically also produce an optimal solution of the problem. This is done by adding suitable pointers such that a simple backtracing starting at an optimal value constructs an optimal solution without increasing the running time.

One of the main drawbacks of dynamic programming algorithms is that they need a lot of space. During the execution of the dynamic programming algorithm above described, for each $i \in \{2, 3, \ldots, n\}$ and $j \in \{1, 2, \ldots, n-1\}$, we have to keep all the values $OPT[S, c_i]$ for all sets of size j and $j+1$. Hence the space needed is $\Omega(2^n)$, which means that not only the running time but also the space used by the algorithm is exponential.

Dynamic Programming is one of the major techniques to design and analyse exact exponential time algorithms. Chapter 3 is dedicated to Dynamic Programming. The relation of exponential space and polynomial space is studied in Chap. 10.

1.3 A Branching Algorithm for Independent Set

A fundamental and powerful technique to design fast exponential time algorithms is Branch & Reduce. It actually comes with many different names: branching algorithm, search tree algorithm, backtracking algorithm, Davis-Putnam type algorithm etc. We shall introduce some of the underlying ideas of the Branch & Reduce paradigm by means of a simple example.

Maximum Independent Set. In the MAXIMUM INDEPENDENT SET problem (MIS), we are given an undirected graph $G = (V, E)$. The task is to find an independent set $I \subseteq V$, i.e. any pair of vertices of I is non-adjacent, of maximum cardinality. For readers unfamiliar with terms from Graph Theory, we provide the most fundamental graph notions in Appendix .

A trivial algorithm for this problem would be to try all possible vertex subsets of G, and for each subset to check (which can be easily done in polynomial time), whether this subset is an independent set. At the end this algorithm outputs the size of the maximum independent set or a maximum independent set found. Since the number of vertex subsets in a graph on n vertices is 2^n, the naive approach here requires time $\Omega(2^n)$.

Here we present a simple branching algorithm for MIS to introduce some of the major ideas. The algorithm is based on the following observations. If a vertex v is in an independent set I, then none of its neighbors can be in I. On the other hand, if I is a maximum (and thus maximal) independent set, and thus if v is not in I then at least one of its neighbors is in I. This is because otherwise $I \cup \{v\}$ would be an independent set, which contradicts the maximality of I. Thus for every vertex v and every maximal independent set I, there is a vertex y from the closed neighborhood $N[v]$ of v, which is the set consisting of v and vertices adjacent to v, such that y is in I, and no other vertex from $N[y]$ is in I. Therefore to solve the problem on G, we solve problems on different reduced instances, and then pick up the best of the obtained solutions. We will refer to this process as branching.

The algorithm in Fig. 1.2 exploits this idea. We pick a vertex of minimum degree and for each vertex from its closed neighborhood we consider a subproblem, where we assume that this vertex belongs to a maximum independent set.

Algorithm mis1(G).
Input: Graph $G = (V, E)$.
Output: The maximum cardinality of an independent set of G.

> **if** $|V| = 0$ **then**
> ⌊ **return** 0
> choose a vertex v of minimum degree in G
> **return** $1 + \max\{\text{mis1}(G \setminus N[y]) : y \in N[v]\}$

Fig. 1.2 Algorithm mis1 for MAXIMUM INDEPENDENT SET

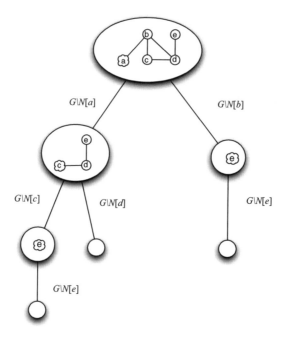

Fig. 1.3 Example of a minimum degree branching algorithm. We branch on vertex a. Then in one subproblem we branch on c, and in the other on e, etc.

The correctness of branching algorithms is usually easy to verify. The algorithm consists of a single branching rule and its correctness follows from the discussions above.

As an example, let us consider the performance of algorithm `mis1` on the graph G of Fig. 1.3. At the beginning the minimum vertex degree is 1, so we select one of the vertices of degree 1, say a. We branch with 2 subproblems, the left branch corresponding to $G \setminus N[a]$ and the right branch to $G \setminus N[b]$. For the right branch there is a unique choice and after branching on e we obtain an empty graph and do not branch anymore. The value the algorithm outputs for this branch is 2 and this corresponds to the maximal independent set $\{b, e\}$. For the left branch we pick a vertex of minimum degree (again 1), say c, and branch again with 2 subproblems. The maximal independent sets found in the left branch are $\{e, c, a\}$ and $\{d, a\}$ and the algorithm reports that the size of a maximum independent set is 3. Let us observe the interesting fact that every maximal independent set can be constructed by following a path from some leaf to the root of the search tree.

Analysing the worst case running time of a branching algorithm can be non-trivial. The main idea is that such an algorithm is recursive and that each execution of it can be seen as a search tree T, where a subproblem, here $G' = G \setminus V'$, is assigned to a node of T. Furthermore when branching from a subproblem assigned to a node of T then any subproblem obtained is assigned to a child of this node. Thus a solution in a node can be obtained from its descendant branches, and this is why we use the term branching for this type of algorithms and call the general approach *Branch & Reduce*. The running time spent by the algorithm on computations corresponding to each node is polynomial—we construct a new graph by removing some vertices, and up to a polynomial multiplicative factor the running time of the algorithm is upper bounded by the number of nodes in the search tree T. Thus to determine the worst case running time of the algorithm, we have to determine the largest number $T(n)$ of nodes in a search tree obtained by any execution of the algorithm on an input graph G having n vertices. To compute $T(n)$ of a branching algorithm one usually relies on the help of linear recurrences. We will discuss in more details how to analyze the running time of such algorithms in Chap. 2.

Let us consider the branching algorithm mis1 for MIS of Fig. 1.2. Let G be the input graph of a subproblem. Suppose the algorithm branches on a vertex v of degree $d(v)$ in G. Let $v_1, v_2, \ldots, v_{d(v)}$ be the neighbors of v in G. Thus for solving the subproblem G the algorithm recursively solves the subproblems $G \setminus N[v]$, $G \setminus N[v_1], \ldots, G \setminus N[v_{d(v)}]$ and we obtain the recurrence

$$T(n) \leq 1 + T(n - d(v) - 1) + \sum_{i=1}^{d(v)} T(n - d(v_i) - 1).$$

Since in step 3 the algorithm chooses a vertex v of minimum degree, we have that for all $i \in \{1, 2, \ldots, d(v)\}$,

$$d(v) \leq d(v_i),$$
$$n - d(v_i) - 1 \leq n - d(v) - 1$$

and, by the monotonicity of $T(n)$,

$$T(n - d(v_i) - 1) \leq T(n - d(v) - 1).$$

We also assume that $T(0) = 1$. Consequently,

$$T(n) \leq 1 + T(n - d(v) - 1) + \sum_{i=1}^{d(v)} T(n - d(v) - 1)$$
$$\leq 1 + (d(v) + 1) \cdot T(n - d(v) - 1).$$

By putting $s = d(v) + 1$, we obtain

$$T(n) \leq 1 + s \cdot T(n - s) \leq 1 + s + s^2 + \cdots + s^{n/s}$$
$$= \frac{1 - s^{n/s+1}}{1 - s} = \mathcal{O}^*(s^{n/s}).$$

For $s > 0$, the function $f(s) = s^{1/s}$ has its maximum value for $s = e$ and for integer s the maximum value of $f(s) = s^{1/s}$ is when $s = 3$.

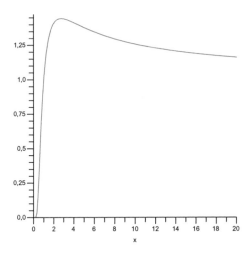

Fig. 1.4 $f(s) = s^{1/s}$

Thus we obtain

$$T(n) = \mathcal{O}^*(3^{n/3}),$$

and hence the running time of the branching algorithm is $\mathcal{O}^*(3^{n/3})$.

Branch & Reduce is one of the fundamental paradigms in the design and analysis of exact exponential time algorithms. We provide a more detailed study of this approach in Chaps 2 and 6.

Notes

As a mathematical problem, TSP was first formulated in 1930 but the history of the problem dates back in the 1800s when Hamilton studied related problems. See [115] for the history of the problem. The dynamic programming algorithm for TSP is due to Bellman [16, 17] and to Held and Karp [111]. Surprisingly, for almost 50 years of developments in Algorithms, the running time $\mathcal{O}^*(2^n)$ of an exact algorithm for TSP has not been improved. Another interesting question is on the space requirements of the algorithm. If the maximum distance between two cities is W, then by making use of inclusion-exclusion (we discuss this technique in Chap. 4), it is possible to solve the problem in time $\mathcal{O}^*(W2^n)$ and space $\mathcal{O}^*(W)$ [127]. Recently, Lokshtanov and Nederlof used the discrete Fourier transform to solve TSP in time $\mathcal{O}^*(W2^n)$ and polynomial, i.e. $n^{\mathcal{O}(1)} \cdot (\log W)^{\mathcal{O}(1)}$ space [154]. See also Chap. 10 for a $\mathcal{O}^*(4^n n^{\mathcal{O}(\log n)})$ time and polynomial space algorithm.

For discussions on computational models we refer to the book of Cormen et al. [52]; see also [61]. The classical algorithm of Schönhage and Strassen from 1971 multiplies two b-bit integers in time $\mathcal{O}(b \log b \log \log b)$ [198]. Recently Fürer improved the running time to $b \log b \, 2^{\mathcal{O}(\log^* b)}$, where $\log^* b$ is the iterated logarithm of b, i.e. the number of times the logarithm function must be iteratively applied before the result is at most 1 [98].

MIS is one of the benchmark problems in exact algorithms. From an exact point of view MIS is equivalent to the problems MAXIMUM CLIQUE and MINIMUM VERTEX COVER. It is easy to modify the branching algorithm mis1 so that it not only finds one maximum independent set but outputs all maximal independent sets of the input graph in time $\mathcal{O}^*(3^{n/3})$. The idea of algorithm mis1 (in a different form) goes back to the works of Miller and Muller [155] from 1960 and to Moon and Moser [161] from 1965 who independently obtained the following combinatorial bound on the maximum number of maximal independent sets.

Theorem 1.1. *The number of maximal independent sets in a graph on n vertices is at most*

$$
\begin{cases}
3^{n/3} & \text{if } n \equiv 0 \pmod 3, \\
4 \cdot 3^{(n-4)/3} & \text{if } n \equiv 1 \pmod 3, \\
2 \cdot 3^{(n-2)/3} & \text{if } n \equiv 2 \pmod 3.
\end{cases}
$$

Moreover, all bounds of Theorem 1.1 are tight and are achievable on graphs consisting of $n/3$ disjoint copies of K_3s; one K_4 or two K_2s and $(n-4)/3$ K_3s; one K_2 and $(n-2)/3$ copies of K_3s. A generalization of this theorem for induced regular subgraphs is discussed in [107].

While the bound $3^{n/3}$ on the number of maximal independent sets is tight, the running time of an algorithm computing a maximum independent set can be strongly improved. The first improvement over $\mathcal{O}^*(3^{n/3})$ was published in 1977 by Tarjan and Trojanowski [213]. It is a Branch & Reduce algorithm of running time $\mathcal{O}^*(2^{n/3}) = \mathcal{O}(1.26^n)$ [213]. In 1986 Jian published an improved algorithm with running time $\mathcal{O}(1.2346^n)$ [125]. In the same year Robson provided an algorithm of running time $\mathcal{O}(1.2278^n)$ [185]. All these three algorithms are Branch & Reduce algorithms, and use polynomial space. In [185] Robson also showed how to speed up Branch & Reduce algorithms using a technique that is now called Memorization (and studied in detail in Chap. 10), and he established an $\mathcal{O}(1.2109^n)$ time algorithm that needs exponential space. Fomin, Grandoni, and Kratsch [85] showed how to solve the problem MIS in time $\mathcal{O}(1.2202^n)$ and polynomial space. Kneis, Langer, and Rossmanith in [133] provided a branching algorithm with a computer-aided case analysis to establish a running time of $\mathcal{O}(1.2132^n)$. Very recently Bourgeois, Escoffier, Paschos and van Rooij in [38] improved the best running time of a polynomial space algorithm to compute a maximum independent set to $\mathcal{O}(1.2114^n)$. A significant amount of research has also been devoted to solving the maximum independent set problem on sparse graphs [13, 37, 39, 47, 48, 97, 179].

Chapter 2
Branching

Branching is one of the basic algorithmic techniques for designing fast exponential time algorithms. It is safe to say that at least half of the published fast exponential time algorithms are branching algorithms. Furthermore, for many NP-hard problems the fastest known exact algorithm is a branching algorithm. Many of those algorithms have been developed during the last ten years by applying techniques like Measure & Conquer, quasiconvex analysis and related ones.

Compared to some other techniques for the design of exact exponential time algorithms, branching algorithms have some nice properties. Typically branching algorithms need only polynomial (or linear) space. The running time on some particular inputs might be much better than the worst case running time. They allow various natural improvements that do not really change the worst case running time but significantly speed up the running time on many instances.

Branching is a fundamental algorithmic technique: a problem is solved by decomposing it into subproblems, recursively solving the subproblems and by finally combining their solutions into a solution of the original problem.

The idea behind branching is so natural and simple that it was reinvented in many areas under different names. In this book we will refer to the paradigm as *Branch & Reduce* and to such algorithms as *branching algorithms*. There are various other names in the literature for such algorithms, like splitting algorithms, backtracking algorithms, search tree algorithms, pruning search trees, DPLL algorithms etc. A branching algorithm is recursively applied to a problem instance and uses two types of rules.

- A *reduction rule* is used to simplify a problem instance or to halt the algorithm.
- A *branching rule* is used to solve a problem instance by recursively solving smaller instances of the problem.

By listing its branching and reduction rules such an algorithm is in principle easy to describe, and its correctness is often quite easy to prove. The crux is the analysis of the worst-case running time.

The design and analysis of such algorithms will be studied in two chapters. In this chapter we start with several branching algorithms and analyze their running

F.V. Fomin, D. Kratsch, *Exact Exponential Algorithms*, Texts in Theoretical
Computer Science. An EATCS Series, DOI 10.1007/978-3-642-16533-7_2,

time by making use of a simple measure of the input, like the number of variables in a Boolean formulae, or the number of vertices in a graph. More involved techniques, which use more complicated measures and allow a better analysis of branching algorithms, in particular Measure & Conquer, will be discussed in Chap. 6.

2.1 Fundamentals

The goal of this section is to present the fundamental ideas and techniques in the design and analysis of branching algorithms; in particular an introduction to the running time analysis of a branching algorithm, including the tools and rules used in such an analysis.

A typical branching algorithm consists of a collection of branching and reduction rules. Such an algorithm may also have (usually trivial) halting rules. Furthermore it has to specify which rule to apply on a particular instance. Typical examples are preference rules or rules on which vertex to branch; e.g. the minimum degree rule of algorithm mis1 in Chap. 1. To a large part designing a branching algorithm means establishing reduction and branching rules.

In many branching algorithms any instance of a subproblem either contains a corresponding partial solution explicitly or such a partial solution can easily be attached to the instance. Thus a given algorithm computing (only) the optimal size of a solution can easily be modified such that it also provides a solution of optimal size. For example, in a branching algorithm computing the maximum cardinality of an independent set in a graph G, it is easy to attach the set of vertices already chosen to be in the (maximum) independent set to the instance.

The correctness of a well constructed branching algorithm usually follows from the fact that the branching algorithm considers all cases that need to be considered. A typical argument is that at least one optimal solution cannot be overlooked by the algorithm. Formally one has to show that all reduction rules and all branching rules are correct. Often this is not even explicitly stated since it is straightforward. Clearly sophisticated rules may need a correctness proof.

Finally let us consider the running time analysis. Search trees are very useful to illustrate an execution of a branching algorithm and to facilitate our understanding of the time analysis of a branching algorithm. A *search tree* of an execution of a branching algorithm is obtained as follows: assign the root node of the search tree to the input of the problem; recursively assign a child to a node for each smaller instance reached by applying a branching rule to the instance of the node. Note that we do not assign a child to a node when a reduction rule is applied. Hence as long as the algorithm applies reduction rules to an instance the instance is simplified but the instance corresponds to the same node of the search tree.

What is the running time of a particular execution of the algorithm on an input instance? To obtain an easy answer, we assume that the running time of the algorithm corresponding to one node of the search tree is polynomial. This has to be guaranteed even in the case that many reduction rules are applied consecutively to one instance before the next branching. Furthermore we require that a reduction of an instance does not produce a simplified instance of larger size. For example, a reduction rule applied to a graph typically generates a graph with fewer vertices.

Under this assumption, which is satisfied for all our branching algorithms, the running time of an execution is equal to the number of nodes of the search tree times a polynomial. Thus analyzing the worst-case running time of a branching algorithm means determining the maximum number of nodes in a search tree corresponding to the execution of the algorithm on an input of size n, where n is typically *not* the length of the input but a natural parameter such as the number of vertices of a graph.

How can we determine the worst-case running time of a branching algorithm? A typical branching algorithm has a running time of $\mathcal{O}^*(\alpha^n)$ for some real constant $\alpha \geq 1$. However except for some very particular branching algorithms we are not able determine the smallest possible α. More precisely, so far no general method to determine the *worst-case* running time of a branching algorithm is available, not even up to a polynomial factor. In fact this is a major open problem of the field. Hence analysing the running time means upper bounding the unknown smallest possible value of α. We shall describe in the sequel how this can be done.

The time analysis of branching algorithms is based on upper bounding the number of nodes of the search tree of any input of size n; and since the number of leaves is at least one half of the number of nodes in any search tree, one usually prefers to upper bound the number of leaves. First a measure for the size of an instance of the problem is defined. In this chapter we mainly consider simple and natural measures like the number of vertices for graphs, the number of variables for Boolean formulas, the number of edges for hypergraphs (or set systems), etc.

We shall see later that other choices of the measure of the size of an instance are possible and useful. The overall approach to analyzing the running time of a branching algorithm that we are going to describe will also work for such measures. This will mainly be discussed in Chap. 6.

Let $T(n)$ be the maximum number of leaves in any search tree of an input of size n when executing a certain branching algorithm. The general approach is to analyse each branching rule separately and finally to use the worst-case time over all branching rules as an upper bound on the running time of the algorithm.

Let b be any branching rule of the algorithm to be analysed. Consider an application of b to any instance of size n. Let $r \geq 2$, and $t_i > 0$ for all $i \in \{1, 2, \ldots, r\}$. Suppose rule b branches the current instance into $r \geq 2$ instances of size at most $n - t_1, n - t_2, \ldots, n - t_r$, for all instances of size $n \geq \max\{t_i : i = 1, 2, \ldots r\}$. Then we call $\mathbf{b} = (t_1, t_2, \ldots t_r)$ the *branching vector* of branching rule b. This implies the linear recurrence

$$T(n) \leq T(n - t_1) + T(n - t_2) + \cdots + T(n - t_r). \tag{2.1}$$

There are well-known standard techniques to solve linear recurrences. A fundamental fact is that base solutions of linear recurrences are of the form c^n for some complex number c, and that a solution of a homogeneous linear recurrence is a linear combination of base solutions. More precisely, a base solution of the linear recurrence (2.1) is of the form $T(n) = c^n$ where c is a complex root of

$$x^n - x^{n-t_1} - x^{n-t_2} - \cdots - x^{n-t_r} = 0. \tag{2.2}$$

We provide some references to the literature on linear recurrences in the Notes.

Clearly worst-case running time analysis is interested in a largest solution of (2.1) which is (up to a polynomial factor) a largest base solution of (2.1). Fortunately there is some very helpful knowledge about the largest solution of a linear recurrence obtained by analysing a branching rule.

Theorem 2.1. *Let b be a branching rule with branching vector $(t_1, t_2, \ldots t_r)$. Then the running time of the branching algorithm using only branching rule b is $\mathcal{O}^*(\alpha^n)$, where α is the unique positive real root of*

$$x^n - x^{n-t_1} - x^{n-t_2} - \cdots - x^{n-t_r} = 0.$$

We call this unique positive real root α the *branching factor* of the branching vector **b**. (They are also called branching numbers.) We denote the branching factor of $(t_1, t_2, \ldots, t_r)$ by $\tau(t_1, t_2, \ldots, t_r)$.

Theorem 2.1 supports the following approach when analysing the running time of a branching algorithm. First compute the branching factor α_i for every branching rule b_i of the branching algorithm to be analysed, as previously described. Now an upper bound of the running time of the branching algorithm is obtained by taking $\alpha = \max_i \alpha_i$. Then the number of leaves of the search tree for an execution of the algorithm on an input of size n is $\mathcal{O}^*(\alpha^n)$, and thus the running time of the branching algorithm is $\mathcal{O}^*(\alpha^n)$.

Suppose a running time expressed in terms of n is what we are interested in. Using n as a simple measure of the size of instances we establish the desired upper bound of the worst-case running time of the algorithm. Using a more complex measure we need to transform the running time in terms of this measure into one in terms of n. This will be discussed in Chap. 6.

Due to the approach described above establishing running times of branching algorithms, they are of the form $\mathcal{O}^*(\alpha^n)$ where $\alpha \geq 1$ is a branching factor and thus a real (typically irrational). Hence the analysis of branching algorithms needs to deal with reals. We adopt the following convention. Reals are represented by five-digit numbers, like 1.3476, such that the real is rounded appropriately: rounded up in running time analysis (and rounded down only in Sect. 6.4). For example, instead of writing $\tau(2,2) = \sqrt{2} = 1.414213562..$, we shall write $\tau(2,2) < 1.4143$, and this implicitly indicates that $1.4142 < \tau(2,2) < 1.4143$. In general, reals are rounded (typically rounded up) and given with a precision of 0.0001.

As a consequence, the running times of branching algorithms in Chap. 2 and Chap. 6 are usually of the form $\mathcal{O}(\beta^n)$, where β is a five-digit number achieved by rounding up a real. See also the preliminaries on the $\mathcal{O}^*$ notation in Chap. 1.

Properties of branching vectors and branching factors are crucial for the design and analysis of branching algorithms. Some of the fundamental ones will be discussed in the remainder of this section.

We start with some easy properties which follow immediately from the definition.

Lemma 2.2. *Let* $r \geq 2$. *Let* $t_i > 0$ *for all* $i \in \{1, 2, \ldots r\}$. *Then the following properties are satisfied:*

1. $\tau(t_1, t_2, \ldots, t_r) > 1$.
2. $\tau(t_1, t_2, \ldots, t_r) = \tau(t_{\pi(1)}, t_{\pi(2)}, \ldots, t_{\pi(r)})$ *for any permutation* π.
3. *If* $t_1 > t_1'$ *then* $\tau(t_1, t_2, \ldots, t_r) < \tau(t_1', t_2, \ldots, t_r)$.

Often the number of potential branching vectors is unbounded, and then repeated use of property 3. of the previous lemma may remove branchings such that a bounded number of branchings remains and the worst-case branching factor remains the same. This is based on the fact that the branching with vector $(t_1', t_2, \ldots, t_r)$ is worse than the branching with vector $(t_1, t_2, \ldots, t_r)$, thus the latter branching can be discarded. We shall also say that $(t_1', t_2, \ldots, t_r)$ *dominates* $(t_1, t_2, \ldots, t_r)$. In general, dominated branching vectors can be safely discarded from a system of branching vectors.

One often has to deal with branching into two subproblems. The corresponding branching vector (t_1, t_2) is called *binary*. Intuitively the factor of a branching vector can be decreased if it is better balanced in the sense of the following lemma.

Lemma 2.3. *Let* i, j, k *be positive reals.*

1. $\tau(k, k) \leq \tau(i, j)$ *for all branching vectors* (i, j) *satisfying* $i + j = 2k$.
2. $\tau(i, j) > \tau(i + \varepsilon, j - \varepsilon)$ *for all* $0 < i < j$ *and all* $0 < \varepsilon < \frac{j-i}{2}$.

The following example shows that trying to improve the balancedness of a branching rule or branching vector by changes in the algorithm or other means is a powerful strategy.

$$\tau(3,3) = \sqrt[3]{2} \quad < 1.2600$$
$$\tau(2,4) = \tau(4,2) < 1.2721$$
$$\tau(1,5) = \tau(5,1) < 1.3248$$

2 A table of precalculated factors of branching vectors can be helpful, in particular in the case of simple analysis or more generally when all branching vectors are integral. Note that the factors have been rounded up, as it is necessary in (upper bounds of) running times. Table 2.1 provides values of $\tau(i, j)$ for all $i, j \in \{1, 2, 3, 4, 5, 6\}$. They have been computed based on the linear recurrence

$$T(n) \leq T(n-i) + T(n-j) \quad \Rightarrow \quad x^n = x^{n-i} + x^{n-j}$$

and the roots of the characteristic polynomial (assuming $(j \geq i)$

$$x^j - x^{j-i} - 1.$$

	1	2	3	4	5	6
1	2.0000	1.6181	1.4656	1.3803	1.3248	1.2852
2	1.6181	1.4143	1.3248	1.2721	1.2366	1.2107
3	1.4656	1.3248	1.2560	1.2208	1.1939	1.1740
4	1.3803	1.2721	1.2208	1.1893	1.1674	1.1510
5	1.3248	1.2366	1.1939	1.1674	1.1487	1.1348
6	1.2852	1.2107	1.1740	1.1510	1.1348	1.1225

Table 2.1 A table of branching factors (rounded up)

So far we have studied fundamental properties of branching vectors. Now we shall discuss a way to combine branching vectors. This is an important tool when designing branching algorithms using simple analysis. We call this operation *addition of branching vectors*.

The motivation is easiest to understand in the following setting. Suppose (i, j) is the branching vector with the largest branching factor of our algorithm. In simple time analysis this means that $\tau(i, j)$ is actually the base of the running time of the algorithm. We might improve upon this running time by modifying our algorithm such that whenever branching with branching vector (i, j) is applied then in one or both of the subproblems obtained, the algorithm branches with a better factor. Thus we would like to know the overall branching factor when executing the branchings together. The hope is to replace the tight factor $\tau(i, j)$ by some smaller value.

The corresponding sum of branching vectors is easy to compute by the use of a search tree; one simply computes the decrease (or a lower bound of the decrease) for all subproblems obtained by the consecutive branchings, i.e. the overall decrease for each leaf of the search tree, and this gives the branching vector for the sum of the branchings. Given that the branching vectors are known for all branching rules applied, it is fairly easy to compute the branching vector of the sum of branchings.

Let us consider an example. Suppose whenever our algorithm (i, j)-branches, i.e. applies a branching rule with branching vector (i, j), it immediately (k, l)-branches on the left subproblem. Then the overall branching vector is $(i + k, i + l, j)$, since the addition of the two branchings produces three subproblems with the corresponding decreases of the size of the original instance. How to find the sum of the branchings by use of a search tree is illustrated in Fig. 2.1. In the second example the branching vectors are $(2, 2)$ and then $(2, 3)$ on the left subproblem and $(1, 4)$ on the right subproblem; the sum is $(4, 5, 3, 6)$.

In Sect. 2.3 we use addition of branching vectors in the design and analysis of an algorithm for MAXIMUM INDEPENDENT SET.

2.2 k-Satisfiability

In this section we describe and analyse a classical branching algorithm solving the k-SATISFIABILITY problem. In 1985 in their milestone paper Monien and Speckenmeyer presented a branching algorithm and its worst-case running time analysis [159]. Their paper is best known for the fact that their algorithm has time complexity $\mathcal{O}(1.6181^n)$ for the 3-SATISFIABILITY problem.

Let us recall some fundamental notions. Let $X = \{x_1, x_2, \ldots, x_n\}$ be a set of *Boolean variables*. A variable or a negated variable is called a *literal*. Let $L = L(X) = \{x_1, \overline{x_1}, x_2, \overline{x_2}, \ldots, x_n, \overline{x_n}\}$ be the set of literals over X. A disjunction $c = (\ell_1 \vee \ell_2 \vee \cdots \vee \ell_t)$ of literals $\ell_i \in L(X)$, $i \in \{1, 2, \ldots t\}$, is called a *clause* over X. As usual we demand that a literal appears at most once in a clause, and that a clause does not contain both a variable x_i and its negation $\overline{x_i}$. We represent a clause c by the set $\{\ell_1, \ell_2, \ldots \ell_t\}$ of its literals. A conjunction $F = (c_1 \wedge c_2 \wedge \cdots \wedge c_r)$ of clauses is called a Boolean formula in *conjunctive normal form* (CNF). We represent F by the set $\{c_1, c_2, \ldots, c_m\}$ of its clauses and call it a *CNF formula*. If each clause of a CNF formula consists of at most k literals then it is called a *k-CNF formula*. By $\emptyset$ we denote the empty formula which is a tautology, and thus satisfiable, by definition. A CNF formula F' is a subformula of a CNF formula F if every clause of F' is a subset of some clause of F.

A *truth assignment t* from X to $\{0, 1\}$ assigns Boolean values (0=false, 1=true) to the variables of X, and thus also to the literals of $L(X)$. A CNF formula F is *satisfiable* if there is a truth assignment t such that the formula F evaluates to true, i.e. every clause contains at least one literal which is true. For example, the CNF formula

$$(x_1 \vee x_2 \vee \overline{x_3}) \wedge (\overline{x_1} \vee \overline{x_2}) \wedge (\overline{x_1} \vee \overline{x_3})$$

is satisfied by the truth assignment $x_1 = $ true, $x_2 = $ false, and $x_3 = $ false.

Satisfiability Problem. In the SATISFIABILITY problem (SAT), we are given a Boolean formula in conjunctive normal form (CNF) with n variables and m clauses. The task is to decide whether the formula is satisfiable, i.e. whether there is a truth assignment for which the formula is true.

k-Satisfiability Problem. If the input of the satisfiability problem is in CNF in which each clause contains at most k literals, then the problem is called the *k*-SATISFIABILITY problem (*k*-SAT).

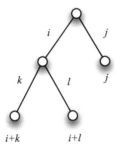
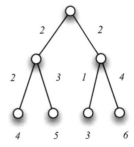

Fig. 2.1 Adding branching vectors

It is well-known that 2-SAT is linear time solvable, while for every $k \geq 3$, k-SAT is NP-complete. How can one decide whether a CNF formula is satisfiable? A trivial algorithm solves SAT by checking every truth assignment of the given CNF formula. If the number of variables is n, then there are 2^n different truth assignments. For each such assignment, we have to check whether every clause is satisfied, which can be done in time $\mathcal{O}(nm)$. Thus the brute-force algorithm will solve the problem in time $\mathcal{O}(2^n nm)$. It is a major open question whether SAT can be solved in time $\mathcal{O}^*((2-\varepsilon)^n)$ for some constant $\varepsilon > 0$.

Now we decribe the branching algorithm of Monien and Speckenmeyer that solves the k-SAT problem in time $\mathcal{O}^*(\alpha_k{}^n)$ where $\alpha_k < 2$ for every fixed $k \geq 3$. Let F be a CNF formula with n variables and m clauses such that each clause of F contains at most k literals. Assuming that F does not have multiple clauses, we have that $m \leq \sum_{i=0}^{k} \binom{2n}{i}$, since F has at most $2n$ literals. Consequently the number of clauses of F is bounded by a polynomial in n. Let $X = \{x_1, x_2, \ldots, x_n\}$ be the set of variables of F and $L = \{x_1, \overline{x_1}, x_2, \overline{x_2}, \ldots, x_n, \overline{x_n}\}$ be the corresponding literals. If $L' \subseteq L$ we denote $\mathrm{lit}(L') = \{x_i, \overline{x_i} : x_i \in L' \text{ or } \overline{x_i} \in L'\}$. The size of a clause, denoted $|c|$, is the number of literals of c. By the definition of k-SAT, $|c| \leq k$ for all clauses of F.

The algorithm recursively computes CNF formulas obtained by a partial truth assignment of the input k-CNF formula, i.e. by fixing the Boolean value of some variables and literals, respectively, of F. Given any partial truth assignment t of the k-CNF formula F the corresponding k-CNF formula F' is obtained by removing all clauses containing a true literal, and by removing all false literals. Hence the instance of any subproblem generated by the algorithm is a k-CNF formula and the size of a formula is its number of variables (or the number of variables not fixed by the corresponding partial truth assignment).

We first study the branching rule of the algorithm. Let F be any k-CNF formula and let $c = (\ell_1 \vee \ell_2 \vee \cdots \vee \ell_q)$ be any clause of F. Branching on clause c means to branch into the following q subformulas obtained by fixing the Boolean values of some literals as described below:

- F_1 : $\ell_1 = \text{true}$
- F_2 : $\ell_1 = \text{false}, \ell_2 = \text{true}$
- F_3 : $\ell_1 = \text{false}, \ell_2 = \text{false}, \ell_3 = \text{true}$
- $\cdots$ $\cdots$
- F_q : $\ell_1 = \text{false}, \ell_2 = \text{false}, \cdots, \ell_{q-1} = \text{false}, \ell_q = \text{true}$

The branching rule says that F is satisfiable if and only if at least one F_i, $i \in \{1, 2, \ldots, q\}$ is satisfiable, and this is obviously correct. Hence recursively solving all subproblem instances F_i, we can decide whether F is satisfiable. The corresponding branching algorithm k-sat1 is described in Fig. 2.2.

Since the Boolean values of i variables of F are fixed to obtain the instance F_i, $i \in \{1, 2, \ldots, q\}$, the number of (non fixed) variables of F_i is $n - i$. Therefore the branching vector of this rule is $(1, 2, \ldots, q)$. To obtain the branching factor of $(1, 2, \ldots, q)$, as discussed in Sect. 2.1, we solve the linear recurrence

$$T(n) \leq T(n-1) + T(n-2) + \cdots + T(n-q)$$

by computing the unique positive real root of

$$x^q = x^{q-1} - x^{q-2} - \cdots - x - 1,$$

which is equivalent to computing the largest real root of

$$x^{q+1} - 2x^q + 1 = 0.$$

For any clause of size q, we denote the branching factor by β_q. Then, when rounding up, one obtains $\beta_2 < 1.6181$, $\beta_3 < 1.8393$, $\beta_4 < 1.9276$ and $\beta_5 < 1.9660$.

We note that on a clause of size 1, there is only one subproblem and thus this is indeed a reduction rule. By adding some simple halting rules saying that a formula containing an empty clause is unsatisfiable and that the empty formula is satisfiable, we would obtain the first branching algorithm consisting essentially of the above branching rule.

Algorithm k-sat1(F).
Input: A CNF formula F.
Output: Return true if F is satisfiable, otherwise return false.

  if *F contains an empty clause* **then**
     L **return** false
  if *F is an empty formula* **then**
     L **return** true
  choose any clause $c = (\ell_1 \vee \ell_2 \vee \cdots \vee \ell_q)$ of F
  $b_1 = \text{k-sat1}(F[\ell_1 = \text{true}])$
  $b_2 = \text{k-sat1}(F[\ell_1 = \text{false}, \ell_2 = \text{true}])$
  $\cdots \qquad \cdots$
  $b_q = \text{k-sat1}(F[\ell_1 = \text{false}, \ell_2 = \text{false}, \cdots, \ell_{q-1} = \text{false}, \ell_q = \text{true}])$
  return $b_1 \vee b_2 \cdots \vee b_q$

Fig. 2.2 Algorithm k-sat1 for *k*-Satisfiability

Of course, we may also add the reduction rule saying that if the formula is in 2-CNF then a polynomial time algorithm will be used to decide whether it is satisfiable. The running time of such a simple branching algorithm is $\mathcal{O}^*(\beta_k{}^n)$ because for a given *k*-CNF as an input all instances generated by the branching algorithm are also *k*-CNF formulas, and thus every clause the algorithm branches on has size at most k.

Notice that the branching factor β_t depends on the size of the clause c chosen to branch on. Hence it is natural to aim at branching on clauses of minimum size. Thus for every CNF formula being an instance of a subproblem the algorithm chooses a clause of minimum size to branch on.

Suppose we can guarantee that for an input *k*-CNF the algorithm always branches on a clause of size at most $k - 1$ (except possibly the very first branching). Such a branching algorithm would solve k-SAT in time $\mathcal{O}^*(\beta_{k-1}{}^n)$. For example, this algorithm would solve 3-SAT in time $\mathcal{O}(1.6181^n)$.

The tool used to achieve this goal is a logical one. Autarkies are partial (truth) assignments satisfying some subset $F' \subseteq F$ (called an autark subset), while not interacting with the clauses in $F \setminus F'$. In other words, a partial truth assignment t of a CNF formula F is called an *autark* if for every clause c of F for which the Boolean value of at least one literal is set by t, there is a literal ℓ_i of c such that $t(\ell_i) = \text{true}$. For example, for the following formula

$$(x_1 \vee x_2 \vee \overline{x_3}) \wedge (\overline{x_1} \vee \overline{x_2}) \wedge (\overline{x_2} \vee \overline{x_3} \vee x_4) \wedge (x_3 \vee \overline{x_4})$$

the partial assignment $x_1 = \text{true}$, $x_2 = \text{false}$ is an autark.

Hence for an autark assignment t of a CNF formula F all clauses for which at least one literal is set by t are simultaneously satisfied by t. Thus if F' is the CNF formula obtained from F by setting Boolean variables w.r.t. t, then F is satisfiable if and only if F' is satisfiable, because no literal of F' has been set by t. On the other hand, if t is not an autark then there is at least one clause c of F which contains a false literal and only false literals w.r.t. t. Hence the corresponding CNF formula F' contains a clause c' with $|c'| < |c|$. Therefore, when a partial truth assignment of a k-CNF formula F is an autark, the algorithm does not branch and is recursively applied to the k-CNF formula F'. When t is not an autark then there is a clause of size at most $k - 1$ to branch on.

In the branching algorithm $\texttt{k-sat2}$ described in Fig. 2.3 this idea is used as follows. First of all there are two reduction rules for termination: If F contains an empty clause then F is unsatisfiable. If F is an empty formula then F is satisfiable. Then the basic branching rule is adapted as follows. Branch on a clause c of F of minimum size. For each subproblem F_i with corresponding truth assignment t_i verify whether t_i is an autark assignment. If none of the assignments $t_i, i \in \{1, 2, \ldots, q\}$ is an autark, then we branch in each of the subproblems as before. However if there is an autark assignment then we recursively solve only the subproblem F' corresponding to this assignment. Indeed, F is satisfiable if and only if F' is satisfiable. This is now a reduction rule.

To complete the running time analysis we notice that whenever branching on a k-CNF formula F one branches on a clause of size at most $k - 1$, except possibly for the very first one. Hence the corresponding branching vector is $(1, 2, \ldots, q)$ with $q \leq k - 1$. Hence the worst case branching factor when executing the algorithm on an input k-CNF formula is $\alpha_k = \beta_{k-1}$. We conclude with the following theorem.

Theorem 2.4. *There is a branching algorithm solving k-SAT with n variables in time $\mathcal{O}^*(\alpha_k{}^n)$, where α_k is the largest real root of $x^k - 2x^{k-1} + 1 = 0$. In particular, the branching algorithm solves 3-SAT in time $\mathcal{O}(1.6181^n)$.*

The best known algorithms solving 3-SAT use different techniques and are based on local search. Chapter 8 is dedicated to local search and the SATISFIABILITY problem.

Algorithm k-sat2(**F**).
Input: A CNF formula F.
Output: Return true if F is satisfiable, otherwise return false.

  if *F contains an empty clause* **then**
      ∟ **return** false
  if *F is an empty formula* **then**
      ∟ **return** true
  choose a clause $c = (\ell_1 \vee \ell_2 \vee \cdots \vee \ell_q)$ of F of minimum size
  let t_1 be the assignment corresponding to $F_1 = F[\ell_1 = \text{true}]$
  let t_2 be the assignment corresponding to $F_2 = F[\ell_1 = \text{false}, \ell_2 = \text{true}]$
  $\cdots$ $\cdots$
  let t_q be the assignment corresponding to
  $F_q = F[\ell_1 = \text{false}, \ell_2 = \text{false}, \cdots, \ell_{q-1} = \text{false}, \ell_q = \text{true}]$
  if *there is an $i \in \{1, 2, \ldots, q\}$ s.t. t_i is autark for F_i* **then**
    | **return** k-sat2(F_i)
  else
    | $b_1 = $ k-sat2($F[\ell_1 = \text{true}]$)
    | $b_2 = $ k-sat2($F[\ell_1 = \text{false}, \ell_2 = \text{true}]$)
    | $\cdots$ $\cdots$
    | $b_q = $ k-sat2($F[\ell_1 = \text{false}, \ell_2 = \text{false}, \cdots, \ell_{q-1} = \text{false}, \ell_q = \text{true}]$)
    | **if** *at least one b_i is* true **then**
    | | **return** true
    | **else**
      ∟ **return** false

Fig. 2.3 Algorithm k-sat2 for k-SATISFIABILITY

2.3 Independent Set

In this section we present a branching algorithm to compute a maximum independent set of an undirected graph. As in the previous section the running time analysis will be simple in the sense that the size of any subproblem instance is measured by the number of vertices in the graph.

Let us recall that an *independent set* $I \subseteq V$ of a graph $G = (V, E)$ is a subset of vertices such that every pair of vertices of I is non-adjacent in G. We denote by $\alpha(G)$ the maximum size of an independent set of G. The MAXIMUM INDEPENDENT SET problem (MIS), to find an independent set of maximum size, is a well-known NP-hard graph problem. A simple branching algorithm of running time $\mathcal{O}^*(3^{n/3}) = \mathcal{O}(1.4423^n)$ to solve the problem MIS exactly has been presented and analysed in Chap. 1.

Our aim is to present a branching algorithm to compute a maximum independent set that uses various of the fundamental ideas of maximum independent set branching algorithms, is not too complicated to describe and has a reasonably good running time when using simple analysis.

Our algorithm works as follows. Let G be the input graph of the current (sub)problem. The algorithm applies reduction rules whenever possible, thus branching rules are applied only if no reduction rule is applicable to the instance. The reduction rules our algorithm uses are the simplicial and the domination rule explained below.

If the minimum degree of G is at most 3 then the algorithm chooses any vertex v of minimum degree. Otherwise the algorithm chooses a vertex v of maximum degree. Then depending on the degree of v, reduction or branching rules are applied and the corresponding subproblem(s) are solved recursively. When the algorithm puts a vertex v into the solution set, we say that it *selects* v. When the algorithm decides that a vertex is not in the solution set, we say that it *discards* v. Before explaining the algorithm in detail, we describe the main rules the algorithm will apply and prove their correctness.

The first one is a reduction rule called the *domination rule*.

Lemma 2.5. *Let $G = (V,E)$ be a graph, let v and w be adjacent vertices of G such that $N[v] \subseteq N[w]$. Then*

$$\alpha(G) = \alpha(G \setminus w).$$

Proof. We have to prove that G has a maximum independent set not containing w. Let I be a maximum independent set of G such that $w \in I$. Since $w \in I$ no neighbor of v except w belongs to I. Hence $(I \setminus \{w\}) \cup \{v\}$ is an independent set of G, and thus a maximum independent set of $G \setminus w$. □

Now let us study the branching rules of our algorithm. The *standard branching* of a maximum independent set algorithm chooses a vertex v and then either it selects v for the solution and solves MIS recursively on $G \setminus N[v]$, or it discards v from the solution and solves MIS recursively on $G \setminus v$. Hence

$$\alpha(G) = \max(1 + \alpha(G \setminus N[v]), \alpha(G \setminus v)).$$

As already discussed in Chap. 1, this standard branching rule of MIS is correct since for any vertex v, G has a maximum independent set containing v or a maximum independent set not containing v. Furthermore if v is selected for any independent set none of its neighbors can be in this independent set. Obviously the branching vector of standard branching is $(d(v) + 1, 1)$.

The following observation is simple and powerful.

Lemma 2.6. *Let $G = (V,E)$ be a graph and let v be a vertex of G. If no maximum independent set of G contains v then every maximum independent set of G contains at least two vertices of $N(v)$.*

Proof. We assume that every maximum independent set of G is also a maximum independent set of $G \setminus v$. Suppose there is a maximum independent set I of $G \setminus v$ containing at most one vertex of $N(v)$. If I contains no vertex of $N[v]$ then $I \cup \{v\}$ is independent and thus I is not a maximum independent set, which is a contradiction. Otherwise, let $I \cap N(v) = \{w\}$. Then $(I \setminus \{w\}) \cup \{v\}$ is an independent set of G, and thus there is a maximum independent set of G containing v, a contradiction. Consequently, every maximum independent set of G contains at least two vertices of $N(v)$. □

Using Lemma 2.6, standard branching has been refined recently. Let $N^2(v)$ be the set of vertices at distance 2 from v in G, i.e. the set of the neighbors of the neighbors of v, except v itself. A vertex $w \in N^2(v)$ is called a *mirror* of v if $N(v) \setminus N(w)$ is a clique. Calling $M(v)$ the set of mirrors of v in G, the standard branching rule can be refined via mirrors.

Lemma 2.7. *Let $G = (V,E)$ be a graph and v a vertex of G. Then*

$$\alpha(G) = \max(1 + \alpha(G \setminus N[v]), \alpha(G \setminus (M(v) \cup \{v\}))).$$

Proof. If G has a maximum independent set containing v then $\alpha(G) = 1 + \alpha(G \setminus N[v])$ and the lemma is true. Otherwise suppose that no maximum independent set of G contains v. Then by Lemma 2.6, every maximum independent set of G contains at least two vertices of $N(v)$. Let w be any mirror of v. This implies that $N(v) \setminus N(w)$ is a clique, and thus at least one vertex of every maximum independent set of G belongs to $N(w)$. Consequently, no maximum independent set of G contains a $w \in M(v)$, and thus w can be safely discarded. □

We call the corresponding rule the *mirror branching*. Its branching vector is $(d(v) + 1, |M(v)| + 1)$.

Exercise 2.8. Show that the following claim is mistaken: If G has an independent set of size k not containing v then there is an independent set of size k not containing v and all its mirrors.

Lemma 2.6 can also be used to establish another reduction rule that we call the *simplicial rule*.

Lemma 2.9. *Let $G = (V,E)$ be a graph and v be a vertex of G such that $N[v]$ is a clique. Then*
$$\alpha(G) = 1 + \alpha(G \setminus N[v]).$$

Proof. If G has a maximum independent set containing v then the lemma is true. Otherwise, by Lemma 2.6 a maximum independent set must contain two vertices of the clique $N(v)$, which is impossible. □

Sometimes our algorithm uses yet another branching rule. Let $S \subseteq V$ be a (small) separator of the graph G, i.e. $G \setminus S$ is disconnected. Then for any maximum independent set I of G, $I \cap S$ is an independent set of G. Thus we may branch into all possible independent sets of S.

Lemma 2.10. *Let G be a graph, let S be a separator of G and let $\mathcal{I}(S)$ be the set of all subsets of S being an independent set of G. Then*

$$\alpha(G) = \max_{A \in \mathcal{I}(S)} |A| + \alpha(G \setminus (S \cup N[A])).$$

Our algorithm uses the corresponding *separator branching* only under the following circumstances: the separator S is the set $N^2(v)$ and this set is of size at most 2.

Thus the branching is done in at most four subproblems. In each of the recursively solved subproblems the input graph is $G \setminus N^2[v]$ or an induced subgraph of it, where $N^2[v] = N[v] \cup N^2(v)$.

Finally let us mention another useful branching rule to be applied to disconnected graphs.

Lemma 2.11. *Let $G = (V, E)$ be a disconnected graph and $C \subseteq V$ a connected component of G. Then*

$$\alpha(G) = \alpha(G[C]) + \alpha(G \setminus C).$$

A branching algorithm for the MAXIMUM INDEPENDENT SET problem based on the above rules is given in Fig. 2.4.

To analyse algorithm `mis2` let us first assume that the input graph G of the (sub)problem has minimum degree at most 3 and that v is a vertex of minimum degree. If $d(v) \in \{0, 1\}$ then the algorithm does not branch. The reductions are obtained by the simplicial rule.

$d(v) = 0$: then v is selected, i.e. added to the maximum independent set I to be computed, and we recursively call the algorithm for $G \setminus v$.

$d(v) = 1$: then v is selected, and we recursively call the algorithm for $G \setminus N[v]$.

Now let us assume that $d(v) = 2$.

$d(v) = 2$: Let u_1 and u_2 be the neighbors of v in G.

(i) $\{u_1, u_2\} \in E$. Then $N[v]$ is a clique and by the simplicial rule $\alpha(G) = 1 + \alpha(G \setminus N[v])$ and the algorithm is recursively called for $G \setminus N[v]$.

(ii) $\{u_1, u_2\} \notin E$. If $|N^2(v)| = 1$ then the algorithm applies a separator branching on $S = N^2(v) = \{w\}$. The two subproblems are obtained either by selecting v and w and recursively calling the algorithm for $G \setminus (N^2[v] \cup N[w])$, or by selecting u_1 and u_2 and recursively calling the algorithm for $G \setminus N^2[v]$. The branching vector is $(|N^2[v] \cup N[w]|, |N^2[v]|)$, and this is at least $(5, 4)$. If $|N^2(v)| \geq 2$ then the algorithm applies a mirror branching to v. If the algorithm discards v then both u_1 and u_2 are selected by Lemma 2.6. If the algorithm selects v it needs to solve $G \setminus N[v]$ recursively. Hence the branching vector is $(N^2[v], N[v])$ which is at least $(5, 3)$. Thus the worst case for $d(v) = 2$ is obtained by the branching vector $(5, 3)$ and $\tau(5, 3) < 1.1939$.

Now let us consider the case $d(v) = 3$.

$d(v) = 3$: Let u_1, u_2 and u_3 be the neighbors of v in G.

(i) $N(v)$ is an independent set: We assume that no reduction rule can be applied. Hence each u_i, $i \in \{1, 2, 3\}$ has a neighbor in $N^2(v)$, otherwise the domination rule could be applied to v and u_i. Note that every $w \in N^2(v)$ with more than one neighbor in $N(v)$ is a mirror of v.

If v has a mirror then the algorithm applies mirror branching to v with branching vector $(|N[v]|, \{v\} \cup M(v))$ which is at least $(4, 2)$ and $\tau(4, 2) < 1.2721$.

If there is no mirror of v then every vertex of $N^2(v)$ has precisely one neighbor in $N(v)$. Furthermore by the choice of v, every vertex of G has degree at least 3, and thus, for all $i \in \{1, 2, 3\}$, the vertex u_i has at least two neighbors in $N^2(v)$. The

---

Algorithm mis2(G).
Input: A graph $G = (V, E)$.
Output: The maximum cardinality of an independent set of G.

if $|V| = 0$ then
  └ **return** 0
if $\exists v \in V$ *with* $d(v) \leq 1$ then
  └ **return** $1 + \texttt{mis2}(G \setminus N[v])$
if $\exists v \in V$ *with* $d(v) = 2$ then
  │ (let u_1 and u_2 be the neighbors of v)
  │ if $\{u_1, u_2\} \in E$ then
  │   └ **return** $1 + \texttt{mis2}(G \setminus N[v])$
  │
  │ if $\{u_1, u_2\} \notin E$ then
  │   │ if $|N^2(v)| = 1$ then
  │   │   │ (let $N^2(v) = \{w\}$)
  │   │   └ **return** $\max(2 + \texttt{mis2}(G \setminus (N^2[v] \cup N[w])), 2 + \texttt{mis2}(G \setminus N^2[v]))$
  │   │ if $|N^2(v)| > 1$ then
  │   │   └ **return** $\max(\texttt{mis2}(G \setminus N[v]), \texttt{mis2}(G \setminus (M(v) \cup \{v\}))$

if $\exists v \in V$ *with* $d(v) = 3$ then
  │ (let $u_1 u_2$ and u_3 be the neighbors of v)
  │ if $G[N(v)]$ *has no edge* then
  │   │ if v *has a mirror* then
  │   │   └ **return** $\max(1 + \texttt{mis2}(G \setminus N[v]), \texttt{mis2}(G \setminus (M(v) \cup \{v\}))$
  │   │ if v *has no mirror* then
  │   │   │ **return** $\max(1 + \texttt{mis2}(G \setminus N[v]), 2 + \texttt{mis2}(G \setminus N[\{u_1, u_2\}]), 2 + \texttt{mis2}(G \setminus$
  │   │   └ $(N[\{u_1, u_3\}] \cup \{u_2\})), 2 + \texttt{mis2}(G \setminus (N[\{u_2, u_3\}] \cup \{u_1\})))$
  │
  │ if $G[N(v)]$ *has one or two edges* then
  │   └ **return** $\max(1 + \texttt{mis2}(G \setminus N[v]), \texttt{mis2}(G \setminus (M(v) \cup \{v\}))$
  │ if $G[N(v)]$ *has three edges* then
  │   └ **return** $1 + \texttt{mis2}(G \setminus N[v])$

if $\Delta(G) \geq 6$ then
  │ choose a vertex v of maximum degree in G
  └ **return** $\max(1 + \texttt{mis2}(G \setminus N[v]), \texttt{mis2}(G \setminus v))$
if G *is disconnected* then
  │ (let $C \subseteq V$ be a component of G)
  └ **return** $\texttt{mis2}(G[C]) + \texttt{mis2}(G \setminus C)$
if G *is 4 or 5-regular* then
  │ choose any vertex v of G
  └ **return** $\max(1 + \texttt{mis2}(G \setminus N[v]), \texttt{mis2}(G \setminus (M(v) \cup \{v\}))$
if $\Delta(G) = 5$ *and* $\delta(G) = 4$ then
  │ choose adjacent vertices v and w with $d(v) = 5$ and $d(w) = 4$ in G
  │ **return**
  └ $\max(1 + \texttt{mis2}(G \setminus N[v]), 1 + \texttt{mis2}(G \setminus (\{v\} \cup M(v) \cup N[w])), \texttt{mis2}(G \setminus (M(v) \cup \{v, w\})))$

Fig. 2.4 Algorithm mis2 for MAXIMUM INDEPENDENT SET

algorithm branches into four subproblems. It either selects v or when discarding v it inspects all possible cases of choosing at least two vertices of $N(v)$ for the solution:

- select v
- discard v, select u_1, select u_2
- discard v, select u_1, discard u_2, select u_3
- discard v, discard u_1, select u_2, select u_3

The branching vector is at least $(4,7,8,8)$ and $\tau(4,7,8,8) < 1.2406$.

(ii) The graph induced by $N(v)$ contains one edge, say $\{u_1, u_2\} \in E$: By the choice of v, the vertex u_3 has degree at least 3, and thus at least two neighbors in $N^2(v)$. Those neighbors of u_3 are mirrors. The algorithm applies mirror branching to v and the branching factor is at least $(4,3)$ and $\tau(4,3) < 1.2208$.

(iii) The graph induced by $N(v)$ contains two edges, say $\{u_1, u_2\} \in E$ and $\{u_2, u_3\} \in E$. Thus when mirror branching on v either v is selected and $G \setminus N[v]$ solved recursively, or v is discarded and u_1 and u_3 are selected. Hence the branching factor is at least $(4,5)$ and $\tau(4,5) < 1.1674$.

(iv) If $N(v)$ is a clique then apply reduction by simplicial rule.

Summarizing the case $d(v) = 3$, the worst case for $d(v) = 3$ is the branching vector $(4,2)$ with $\tau(4,2) < 1.2721$. It dominates the branching vectors $(4,3)$ and $(4,5)$ which can thus be discarded. There is also the branching vector $(4,7,8,8)$.

Now assume that the input graph G has minimum degree at least 4. Then the algorithm does not choose a minimum degree vertex (as in all cases above). It chooses a vertex v of maximum degree to branch on it.

$\mathbf{d(v) \geq 6}$: The algorithm applies mirror branching to v. Thus the branching vector is $(d(v) + 1, 1)$ which is at least $(7,1)$ and $\tau(7,1) < 1.2554$.

$\mathbf{d(v) = 4}$: Due to the branching rule applied to disconnected graphs the graph G is connected. Furthermore G is 4-regular since its minimum degree is 4.

For any $r \geq 3$, there is at most one r-regular graph assigned to a node of the search tree from the root to a leaf, since every instance generated by the algorithm is an induced subgraph of the input graph of the original problem. Hence any branching rule applied to r-regular graphs, for some fixed r, can only increase the number of leaves by a multiplicative constant. Hence we may neglect the branching rules needed for r-regular graphs in the time analysis.

$\mathbf{d(v) = 5}$: A similar argument applies to 5-regular graphs, and thus we do not have to consider 5-regular graphs. Therefore we may assume that the graph G has vertices of degree 5 and vertices of degree 4, and no others since vertices of smaller or higher degrees would have been chosen to branch on in earlier parts of the algorithm. As G is connected there is a vertex of degree 5 adjacent to a vertex of degree 4. The algorithm chooses those vertices to be v and w such that $d(v) = 5$ and $d(w) = 4$. Now it applies a mirror branching on v. If there is a mirror of v then the branching vector is at least $(2,6)$ and $\tau(2,6) < 1.2366$. If there is no mirror of v then mirror branching on v has branching vector $(1,6)$ and $\tau(1,6) < 1.2852$.

To achieve a better overall running time of the branching algorithm than $\Theta^*(1.2852^n)$ we use addition of branching vectors. Note that the subproblem obtained by discarding v in the mirror branching, i.e. the instance $G \setminus v$, contains the vertex w such that the degree of w in $G \setminus v$ is equal to 3. Now whenever the algorithm does the (expensive) $(1,6)$-branching it immediately branches on w in the subproblem obtained by discarding v. Hence we may replace $(1,6)$ by the branching vectors obtained by addition. The candidates for addition are the branching vectors for branching on a vertex of degree 3. These are $(4,2)$ and $(4,7,8,8)$. Adding $(4,2)$ to $(1,6)$ (in the subproblem achieved by a gain of 1) gives $(5,3,6)$. Adding $(4,7,8,8)$ to $(1,6)$ gives $(5,8,9,9,6)$. The corresponding branching factors are $\tau(5,6,8,9,9) < 1.2548$ and $\tau(3,5,6) < 1.2786$. See Fig. 2.5.

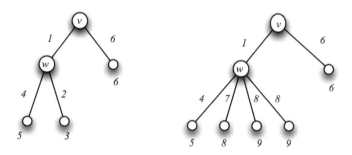

Fig. 2.5 Adding branching vectors in the case $d(v) = 5$

Consequently the worst case branching factor of the whole branching algorithm is $\tau(3,5,6) < 1.2786$.

Theorem 2.12. *The branching algorithm* mis2 *for the* MAXIMUM INDEPENDENT SET *problem has running time* $\mathcal{O}(1.2786^n)$.

Exercise 2.13. Improve upon the branching algorithm mis2 by a more careful analysis of the case $d(v) = 3$.

Exercise 2.14. Improve upon the branching algorithm mis1 of Chap. 1 by the use of Lemma 2.6.

Notes

The analysis of the worst-case running time of branching algorithms is based on solving linear recurrences. This is a subject of most textbooks in Discrete Mathematics. We recommend Rosen [191] and Graham, Knuth, Patashnik [104].

The history of branching algorithms can be traced back to the work of Davis and Putnam [63] (see also [62]) on the design and analysis of algorithms to solve the SATISFIABILITY problem (SAT). It is well-known that SAT and 3-SAT are

NP-complete [100], while there is a linear time algorithm solving 2-SAT [8]. For a few decades branching algorithms have played an important role in the design of exact algorithms for SAT and in particular 3-SAT. One of the milestone branching algorithms for 3-SAT and k-SAT, $k \geq 3$, has been established by Monien and Speckenmeyer in 1985 [159]. In his long paper from 1999 [147], Kullmann establishes a branching algorithm of running time $\mathcal{O}(1.5045^n)$. He also presents a comprehensive study of the worst-case analysis of branching algorithms. See also [148] for an overview. Parts of the Fundamentals section are inspired by [147]. General references for algorithms and complexity of SAT are [21, 130]. Branching algorithms are used to solve different variants of SAT like Max-2-SAT, XSAT , counting variants, and SAT parameterized by the number of clauses [57, 44, 58, 105, 112, 140, 144, 145, 146].

Another problem with a long history in branching algorithms is the MAXIMUM INDEPENDENT SET problem dating back to 1977 and the algorithm of Tarjan and Trojanowski [213]. We refer the reader to the Notes of Chap. 1 for more information.

Many of the reduction and branching rules for MAXIMUM INDEPENDENT SET can be found in [213]. Fürer used the separator rule in [97]. The mirror rule was been introduced by Fomin, Grandoni and Kratsch as part of their Measure & Conquer-based branching algorithm for MAXIMUM INDEPENDENT SET [85], which also uses the folding rule. Kneis, Langer, and Rossmanith in [133] introduced the satellite rule.

Chvátal studied the behaviour of branching algorithms for MIS and showed that for a large class of branching algorithms there is a constant $c > 1$ such that for almost all graphs the running time of these algorithms is more than c^n [50]. Lower bounds for different variants of DPLL algorithms for SAT are discussed in [1, 3, 166].

Chapter 3
Dynamic Programming

Dynamic programming is one of the basic algorithmic techniques. Contrary to branching algorithms, dynamic programming is of great importance for designing polynomial time algorithms as well as for designing exponential time algorithms. The main idea of dynamic programming is to start by solving small or trivial instances and then gradually resolving larger and harder subproblems by composing solutions from smaller subproblems. From this point of view, dynamic programming is quite opposite to branching, where we try to decompose the problem. The careful choice of subproblems is often crucial for the running time of dynamic programming. We refer to classical textbooks [52, 129] for detailed discussions of dynamic programming and its applications in polynomial time algorithms. The (unwanted) property of exponential time dynamic programming algorithms is that they usually require exponential space, contrary to branching algorithms which usually need only polynomial space. The relation of time and space requirements of exponential time algorithms is studied in Chap. 10.

In this chapter we give several examples of the use of dynamic programming to design exact exponential time algorithms. In Chap. 1, we showed how dynamic programming across the subsets is used to solve the TRAVELLING SALESMAN problem on n cities in time $\mathcal{O}(2^n n^2)$. This approach can be applied to many permutation problems. In many examples, the number of subproblems we have to solve is proportional to the number of subsets of the ground set up to a polynomial factor and in such cases dynamic programming requires $\mathcal{O}^*(2^n)$ steps. Sometimes the number of "essential" subsets is significantly smaller than 2^n and in that case faster exponential time algorithms can be established.

F.V. Fomin, D. Kratsch, *Exact Exponential Algorithms*, Texts in Theoretical
Computer Science. An EATCS Series, DOI 10.1007/978-3-642-16533-7_3,
© Springer-Verlag Berlin Heidelberg 2010

3.1 Basic Examples

3.1.1 Permutation Problems

In this subsection we discuss how to use dynamic programming across the subsets to solve different permutation problems. In all these problems one has to find an optimal permutation of the elements of the ground set. One of the typical examples is the TRAVELLING SALESMAN problem (TSP), discussed in Chap. 1, where one asks for a permutation minimizing the total cost of a tour. Most of the permutation problems on a ground set of size n can be solved trivially in time $\mathcal{O}^*(n!)$. In Chap. 1, we showed how dynamic programming across the subsets for TSP results in an algorithm of running time $\mathcal{O}^*(2^n)$. For many other permutation problems, dynamic programming allows us to reduce the running time to $\mathcal{O}(c^n)$, for some constant $c > 1$.

Scheduling Problem. Let us consider a SCHEDULING problem in which we are given a set of jobs $J_1, J_2, \ldots, J_n$, execution times $\tau_1, \tau_2, \ldots, \tau_n$, and cost functions $c_1(t), c_2(t), \ldots, c_n(t)$. Every job J_i requires time τ_i to be executed, and there is an associated cost $c_i(t)$ when finishing job J_i at time t. The jobs have to be executed on a single machine. We assume that the machine is to be in use constantly and that an executed job is not to be interrupted before its completion (i.e. non-preemptive scheduling). Under these assumptions the order of executing jobs, or *schedule*, can be represented by a permutation π of $\{1, 2, \ldots, n\}$. Given such a schedule, the termination time of job $J_{\pi(i)}$ is

$$t_{\pi(i)} = \sum_{j=1}^{i} \tau_{\pi(j)}$$

and the total cost associated with the schedule is

$$\sum_{i=1}^{n} c_{\pi(i)} \cdot t_{\pi(i)}.$$

For a given input, the task is to find the minimum total cost taken over all possible schedules; we call this minimum cost *OPT*.

For a subset of jobs S, we denote the time required to finish all jobs from S by t_S, i.e.

$$t_S = \sum_{i \in S} \tau_i.$$

Let $OPT[S]$ be the minimum cost of executing all jobs of S in the interval $[0, t_S]$. In case $S = \{J_i\}$, we have $OPT[S] = c_i(\tau_i)$. If $|S| > 1$, then

$$OPT[S] = \min\{OPT[S \setminus \{J_i\}] + c_i(t_S)\} : J_i \in S\}. \tag{3.1}$$

To see this, if in some optimal scheduling of S job J_i is the last executed job, then

$$OPT[S] = OPT[S \setminus \{J_i\}] + c_i(t_S).$$

Taking the minimum over all choices of J_i, we arrive at (3.1).

Now OPT can be computed by dynamic programming based on (3.1). The number of steps is $\mathcal{O}(n2^n)$ and we conclude with the following.

Theorem 3.1. *The scheduling problem is solvable in time $\mathcal{O}(n2^n)$.*

There is a natural way to establish an algorithm based on the recurrence achieved. The corresponding algorithm cost is described in Fig. 3.1.

Algorithm cost$(J_1, J_2, \ldots, J_n, \tau_1, \tau_2, \ldots, \tau_n, c_1(t), c_2(t), \ldots, c_n(t))$.
Input: A collection of jobs $J_1, J_2, \ldots, J_n$ with execution times $\tau_1, \tau_2, \ldots, \tau_n$, and cost functions
$c_1(t), c_2(t), \ldots, c_n(t)$.
Output: The minimum cost OPT of a schedule.

   for $i = 1$ *to* n **do**
      $OPT[\{J_i\}] = c_i(\tau_i)$

   for $j = 2$ *to* n **do**
      forall $S \subseteq \{1, 2, 3, \ldots n\}$ with $|S| = j$ **do**
         $OPT[S] = \min\{OPT[S \setminus \{J_i\}] + c_i(t_S) : J_i \in S\}$

   return $OPT[\{1, 2, \ldots, n\}]$

Fig. 3.1 Algorithm cost for the SCHEDULING problem

Exercise 3.2. The *cutwidth* of a vertex ordering π of a graph $G = (V, E)$ is

$$\max_{v \in V} |\{\{w, x\} \in E : \pi(w) \leq \pi(v) < \pi(x)\}|.$$

The cutwidth of a graph $G = (V, E)$ is the minimum cutwidth taken over all linear orderings of its vertices. Prove that the cutwidth of a graph on n vertices can be computed in time $\mathcal{O}^*(2^n)$.

Exercise 3.3. Design and analyse dynamic programming algorithms to compute the pathwidth of a graph. See Chap. 5 for the definitions of pathwidth.

Directed Feedback Arc Set. In the DIRECTED FEEDBACK ARC SET problem we are given a directed graph $G = (V, A)$. The task is to find a feedback arc set of minimum cardinality, i.e. a set of arcs $F \subseteq A$ such that $(V, A \setminus F)$ is acyclic and $|F|$ is as small as possible. Hence $F \subseteq A$ is a feedback arc set of the directed graph $G = (V, A)$ if each directed cycle of G contains at least one arc of F.

At first glance the problem does not look like a permutation problem. However it can be expressed in terms of permutations. Let us recall that a *topological ordering* of a directed graph $G = (V, A)$ is an ordering $\pi : V \to \{1, 2, \ldots, |V|\}$ (or permutation) of its vertices such that all arcs are directed from left to right. In other words, for any arc $(u, v) \in A$, we have $\pi(u) < \pi(v)$.

Lemma 3.4. *Let $G = (V,A)$ be a directed graph, and let $w : A \to \mathbb{N}^+$ be a function assigning to each arc a non-negative integer weight. Let $k \geq 0$ be an integer. There exists a set of arcs $F \subseteq A$ such that $(V, A \setminus F)$ is acyclic and $\sum_{a \in F} w(a) \leq k$, if and only if there is a linear ordering π of G, such that*

$$\sum_{(x,y) \in A, \ \pi(x) > \pi(y)} w((x,y)) \leq k.$$

Proof. The proof of the lemma becomes trivial after we make the following observation: A directed graph is acyclic if and only if it has a topological ordering.

Indeed, if a graph is acyclic, it contains a vertex of in-degree 0. We take this vertex as the first vertex in the ordering, remove it from the graph and proceed recursively. In the opposite direction, if a directed graph has a topological ordering, then it cannot contain a cycle, because in any ordering, at least one arc of the cycle should go from right to left.   □

By making use of Lemma 3.4, it is now easy to obtain a recurrence like (1.1) or (3.1) and to solve DIRECTED FEEDBACK ARC SET in time $\mathcal{O}(nm2^n)$.

Theorem 3.5. *The DIRECTED FEEDBACK ARC SET problem can be solved in time $\mathcal{O}(nm2^n)$, where n is the number of vertices and m is the number of arcs of the given weighted directed graph G.*

Exercise 3.6. We leave the proof of Theorem 3.5 as an exercise for the reader.

Exercise 3.7. In the OPTIMAL LINEAR ARRANGEMENT problem, we are given a graph $G = (V, E)$. The task is to find the minimum value of

$$\sum_{\{v,w\} \in E} |\pi(v) - \pi(w)|,$$

where the minimum is taken over all linear orderings π of the vertices of G. Prove that OPTIMAL LINEAR ARRANGEMENT is solvable in time $\mathcal{O}^*(2^n)$.
Hint: Restate the problem as a permutation problem by proving that for each linear ordering π of V,

$$\sum_{\{v,w\} \in E} |\pi(v) - \pi(w)| = \sum_{v \in V} |\{\{x,y\} \in E \ : \ \pi(x) \leq \pi(v) < \pi(w)\}|.$$

3.1.2 Partition Problems

Graph Coloring. A k-coloring c of an undirected graph $G = (V, E)$ assigns a color to each vertex of the graph $c : V \to \{1, 2, \ldots, k\}$ such that adjacent vertices have different colors. The smallest k for which G has a k-coloring is called the *chromatic number* of G, denoted by $\chi(G)$. A coloring c of G using $\chi(G)$ colors is called an *optimal coloring.* In the COLORING problem we are given an undirected graph $G = (V, E)$. The task is to compute the chromatic number of G, or even to find an optimal coloring of G.

The computation of the chromatic number of a graph is a typical partition problem, and the trivial brute-force solution would be for every vertex to try every possible color. The maximum chromatic number of an n-vertex graph is equal to n; thus such a trivial algorithm has running time $\mathcal{O}^*(n^n)$ which is roughly $2^{\mathcal{O}(n \log n)}$.

Theorem 3.8. *The chromatic number of an n-vertex graph can be computed in time* $\mathcal{O}^*((1 + \sqrt[3]{3})^n) = \mathcal{O}(2.4423^n)$ *by a dynamic programming algorithm.*

Proof. For every $X \subseteq V$, we define $OPT[X] = \chi(G[X])$, the chromatic number of the subgraph of G induced by X. The algorithm computes the values of $OPT[X]$ by making use of dynamic programming. For every subset X in the order of increasing cardinality the following recurrence is used:

$$OPT[\emptyset] = 0,$$

and

$$OPT[X] = 1 + \min\{OPT[X \setminus I] : I \text{ is a maximal independent set of } G[X]\}.$$

We claim that $\chi(G) = OPT[V]$. Indeed, every k-coloring of a graph G is a partition of the vertex set into k independent sets (resp. color classes) and that we may always modify the k-coloring such that one independent set is maximal. Therefore an optimal coloring of G is obtained by removing a maximal independent set I from G and adding an optimal coloring of $G \setminus I$.

What is the running time of this algorithm? Let $n = |V|$. The algorithm runs on all subsets $X \subseteq V$, and for every subset X, it runs on all its subsets I, which are maximal independent sets in $G[X]$. The number of such sets is at most $2^{|X|}$. Thus the number of steps of the algorithm is up to a polynomial factor at most

$$\sum_{i=1}^{n} \binom{n}{i} \cdot 2^i = 3^n. \tag{3.2}$$

In (3.2) we do not take into account that the algorithm does not run on all subsets of a subset X, but only on maximal independent sets of $G[X]$. As we already know from Chap. 1, the number of maximal independent sets in a graph of i vertices is at most $3^{i/3}$ and these sets can be enumerated in time $\mathcal{O}^*(3^{i/3})$. Thus up to a polynomial factor, the running time of the algorithm can be bounded by

$$\sum_{i=1}^{n} \binom{n}{i} \cdot 3^{i/3} = (1 + \sqrt[3]{3})^n < 2.4423^n. \tag{3.3}$$

$\square$

The running time obtained in Theorem 3.8 can be improved to $\mathcal{O}^*(2^n)$ by combining dynamic programming with inclusion-exclusion. We will come back to this in Chap. 4.

Exercise 3.9. In the DOMATIC NUMBER problem we are given an undirected graph $G = (V, E)$. The task is to compute the domatic number of G which is the largest integer k such that there is a partition of V into pairwise disjoint sets $V_1, V_2, \ldots V_k$ such that $V_1 \cup V_2 \cup \cdots \cup V_k = V$ and each V_i is a dominating set of G. Show how to compute the domatic number of an n-vertex graph in time $\mathcal{O}^*(3^n)$.

3.2 Set Cover and Dominating Set

Minimum Set Cover. In the MINIMUM SET COVER problem (MSC) we are given a universe $\mathcal{U}$ of elements and a collection $\mathcal{S}$ of (non-empty) subsets of $\mathcal{U}$. The task is to find the minimum cardinality of a subset $\mathcal{S}' \subseteq \mathcal{S}$ which *covers* $\mathcal{U}$, i.e. $\mathcal{S}'$ satisfies

$$\cup_{S \in \mathcal{S}'} S = \mathcal{U}.$$

A *minimum set cover* of $\mathcal{U}$ is a minimum cardinality subset $\mathcal{S}' \subseteq \mathcal{S}$ which covers $\mathcal{U}$.

Let $\mathcal{U}$ be a ground set of n elements, and let $\mathcal{S} = \{S_1, S_2, \ldots, S_m\}$ be a collection of subsets of $\mathcal{U}$. We say that a subset $\mathcal{S}' \subseteq \mathcal{S}$ *covers* a subset $S \subseteq \mathcal{U}$, if every element in S belongs to at least one member of $\mathcal{S}'$.

Note that a minimum set cover of $(\mathcal{U}, \mathcal{S})$ can trivially be found in time $\mathcal{O}(n2^m)$ by checking all subsets of $\mathcal{S}$.

Theorem 3.10. *There is an $\mathcal{O}(nm2^n)$ time algorithm to solve the MSC problem for any instance $(\mathcal{U}, \mathcal{S})$ where $|\mathcal{U}| = n$ and $|\mathcal{S}| = m$.*

Proof. Let $(\mathcal{U}, \mathcal{S})$ with $\mathcal{S} = \{S_1, S_2, \ldots, S_m\}$ be an instance of the minimum set cover problem over a ground set $\mathcal{U}$ with $|\mathcal{U}| = n$.

For every nonempty subset $U \subseteq \mathcal{U}$, and for every $j = 1, 2, \ldots, m$ we define $OPT[U; j]$ as the minimum cardinality of a subset of $\{S_1, \ldots, S_j\}$ that covers U. If $\{S_1, \ldots, S_j\}$ does not cover U then we set $OPT[U; j] := \infty$.

Now all values $OPT[U; j]$ can be computed as follows. In the first step, for every subset $U \subseteq \mathcal{U}$, we set $OPT[U; 1] = 1$ if $U \subseteq S_1$, and $OPT[U; 1] = \infty$ otherwise. Then in step $j+1$, $j \in \{1, 2, \ldots, m-1\}$, $OPT[U; j+1]$ is computed for all $U \subseteq \mathcal{U}$ in $\mathcal{O}(n)$ time as follows:

$$OPT[U; j+1] = \min\{OPT[U; j], \ OPT[U \setminus S_{j+1}; j] + 1\}.$$

This yields an algorithm to compute $OPT[U; j]$ for all $U \subseteq \mathcal{U}$ and all $j \in \{1, 2, \ldots, m\}$ of overall running time $\mathcal{O}(nm2^n)$. Therefore $OPT[\mathcal{U}; m]$ is the cardinality of a minimum set cover of $(\mathcal{U}, \mathcal{S})$. $\qquad\square$

Now let us show how the dynamic programming algorithm of Theorem 3.10 can be used to *break the 2^n barrier* for the MINIMUM DOMINATING SET problem; that is to construct an $\mathcal{O}^*(c^n)$ algorithm with $c < 2$ solving this problem.

Minimum Dominating Set. In the MINIMUM DOMINATING SET problem (MDS) we are given an undirected graph $G = (V, E)$. The task is to find the minimum cardinality of a dominating set in G.

A vertex subset $D \subseteq V$ of a graph $G = (V, E)$ is a *dominating set* for G if every vertex of G is either in D, or adjacent to some vertex in D. The *domination number* $\gamma(G)$ of a graph G is the minimum cardinality of a dominating set of G. The MDS problem asks us to compute $\gamma(G)$.

The MINIMUM DOMINATING SET problem can be reduced to the MINIMUM SET COVER problem by imposing $\mathcal{U} = V$ and $\mathcal{S} = \{N[v] : v \in V\}$. Note that $N[v]$ is the set of vertices dominated by v, thus D is a dominating set of G if and only if $\{N[v] : v \in D\}$ is a set cover of $\{N[v] : v \in V\}$. In particular, every minimum set cover of $\{N[v] : v \in V\}$ corresponds to a minimum dominating set of G.

We use Theorem 3.10 to prove the following result.

Theorem 3.11. *Let $G = (V, E)$ be a graph on n vertices given with an independent set I. Then a minimum dominating set of G can be computed in time $2^{n-|I|} \cdot n^{\mathcal{O}(1)}$. In particular, a minimum dominating set of a bipartite graph on n vertices can be computed in time $\mathcal{O}^*(2^{n/2})$.*

Proof. Let $J = V \setminus I$ be the set of vertices outside the independent set I. Instead of trying all possible subsets D of V as dominating sets, we try all possible projections of D on J, and for each such projection $J_D = J \cap D$, we decide whether J_D can be extended to a dominating set of G by adding only vertices of I. In fact, for every $J_D \subseteq J$ the smallest possible number of vertices of I should be added to J_D to obtain a dominating set of G

For every subset $J_D \subseteq J$, we show how to construct a set D such that

$$|D| = \min\{|D'| : D' \text{ is a dominating set and } J \cap D' = J_D\}. \qquad (3.4)$$

Then obviously

$$\gamma(G) = \min_{J_D \subseteq J} |D|.$$

The set $I_D = I \setminus N(J_D)$ is a subset of D since I is an independent set and the vertices of I_D cannot be dominated by J_D. Then the only vertices that are not dominated by $I_D \cup J_D$ are the vertices $J_X = J \setminus (N[J_D] \cup N(I_D))$. Therefore, to find D we have to add to $I_D \cup J_D$ the minimum number of vertices from $I \setminus I_D$ that dominate all vertices of J_X. To find a minimum subset of $I \setminus I_D$ which dominates J_X, we reduce this problem to MSC by imposing $\mathcal{U} = J_X$ and $\mathcal{S} = \{N[v] : v \in I \setminus I_D\}$. By Theorem 3.10, such a problem is solvable in time $2^{|J_X|} \cdot n^{\mathcal{O}(1)}$. Thus the running time of the algorithm (up to a polynomial factor) is

$$\sum_{J_D \subseteq J} \binom{|J|}{|J_D|} \cdot 2^{|J_X|} = \sum_{J_D \subseteq J} \binom{|J|}{|J_D|} \cdot 2^{|J \setminus (N[J_D] \cup N(I_D))|}$$

$$\leq \sum_{J_D \subseteq J} \binom{J}{|J_D|} \cdot 2^{|J \setminus J_D|}$$

$$= \sum_{J_D \subseteq J} \binom{n - |I|}{|J_D|} \cdot 2^{n - |I| - |J_D|}$$

$$= 3^{n - |I|}.$$

In the remaining part of the proof we show how to improve the running time $3^{n-|I|}$ to the claimed $2^{n-|I|}$. There is a general theorem based on fast subset convolution which will be proved in Chap. 7 that can do this job. However, for this case, we show an alternative proof based on dynamic programming.

Once again, we want to show that

$$\sum_{J_D \subseteq J} \binom{J}{|J_D|} \cdot 2^{|J \setminus J_D|}$$

can be evaluated in time $2^{n-|I|} \cdot n^{O(1)}$. Instead of trying all subsets of J and then for each subset constructing a dominating set D, we do the following. For every subset $X \subseteq J$ we compute a minimum subset of I which dominates X. We can compute this by performing the following dynamic programming. Let us fix an ordering $\{v_1, v_2, \ldots, v_k\}$ of I. We define $D_{X,i}$ a subset of $\{v_1, v_2, \ldots, v_i\}$ of the minimum size subset which dominates X. Thus $D_{X,k}$ is a subset of I dominating X of minimum size. We put $D_{\emptyset,k} = \emptyset$ and for $X \neq \emptyset$,

$$D_{X,1} = \begin{cases} v_1, & \text{if } X \subseteq N(v_1), \\ \{v_1, v_2, \ldots, v_k\}, & \text{otherwise.} \end{cases}$$

To compute the values $D_{X,i}$ for $i > 1$, we consider two cases. Either the optimum set is a subset of $\{v_1, v_2, \ldots, v_{i-1}\}$, or it contains v_i. Thus

$$D_{X,i} = \begin{cases} D_{X,i-1}, & \text{if } |D_{X,i-1}| < |D_{X \setminus N(v_i), i-1}| + 1, \\ D_{X \setminus N(v_i), i-1} \cup \{v_i\}, & \text{otherwise.} \end{cases}$$

The computation of all sets $D_{X,k}$, $X \subseteq J$, takes time $2^{|J|} \cdot n^{O(1)}$. Once these sets are computed, constructing a set D satisfying (3.4) for every subset $J_D \subseteq J$ can be done in polynomial time by computing $D = I_D \cup J_D \cup D_{J_X,k}$. In total, the running time needed to compute $\gamma(G)$ is the time required to compute sets $D_{X,k}$ plus the time required to compute for every subset J_D subsets I_D and $D_{J_X,k}$. Thus, up to a polynomial factor, the running time is

$$2^{|J|} + \sum_{J_D \subseteq J} \binom{|J|}{|J_D|} = 2^{|J|+1} = 2^{n - |I| + 1}.$$

$\Box$

The following binary entropy function is very helpful in computations involving binomial coefficients. For more information on Stirling's formula and the binary entropy function the reader is referred to the Notes.

Lemma 3.12 (Stirling's Formula). *For $n > 0$,*

$$\sqrt{2\pi n}\left(\frac{n}{e}\right)^n \leq n! \leq 2\sqrt{2\pi n}\left(\frac{n}{e}\right)^n.$$

The *binary entropy function* h is defined by

$$h(\alpha) = -\alpha \log_2 \alpha - (1-\alpha)\log_2(1-\alpha)$$

for $\alpha \in (0,1)$.

Lemma 3.13 (Entropy function). *For $\alpha \in (0,1)$,*

$$\frac{1}{\sqrt{8n\alpha(1-\alpha)}} \cdot 2^{h(\alpha)n} \leq \sum_{i=1}^{\alpha n}\binom{n}{i} \leq 2^{h(\alpha)n} = \left(\frac{1}{\alpha}\right)^{\alpha n} \cdot \left(\frac{1}{1-\alpha}\right)^{(1-\alpha)n}.$$

Proof. We give only the proof of the second inequality.
Since $h(\alpha) = h(1-\alpha)$, we can assume that $\alpha \leq 1/2$. By the Binomial Theorem,

$$\begin{aligned}
1 = (\alpha + (1-\alpha))^n &= \sum_{i=1}^{n}\binom{n}{i}\alpha^i(1-\alpha)^{n-i}\\
&= \sum_{i=1}^{n}\binom{n}{i}\left(\frac{\alpha}{1-\alpha}\right)^i(1-\alpha)^n\\
&\geq \sum_{i=1}^{\alpha n}\binom{n}{i}\left(\frac{\alpha}{1-\alpha}\right)^i(1-\alpha)^n
\end{aligned}$$

Because $0 \leq \alpha \leq 1/2$ and $i \leq \alpha n$, we have that

$$\left(\frac{\alpha}{1-\alpha}\right)^i \geq \left(\frac{\alpha}{1-\alpha}\right)^{\alpha n}.$$

Therefore,

$$\begin{aligned}
1 \geq \sum_{i=1}^{\alpha n}\binom{n}{i}\left(\frac{\alpha}{1-\alpha}\right)^i(1-\alpha)^n &\geq \sum_{i=1}^{\alpha n}\binom{n}{i}\left(\frac{\alpha}{1-\alpha}\right)^{\alpha n}(1-\alpha)^n\\
&= \sum_{i=1}^{\alpha n}\binom{n}{i}\alpha^{\alpha n}(1-\alpha)^{(1-\alpha)n}\\
&= \sum_{i=1}^{\alpha n}\binom{n}{i}2^{-nh(\alpha)}.
\end{aligned}$$

$$\square$$

By making use of Theorem 3.11 and the binary entropy function, it is possible to construct an algorithm solving the MDS problem (for general graphs) faster than by trivial brute-force in $\Theta^*(2^n)$.

Corollary 3.14. *The* MINIMUM DOMINATING SET *problem is solvable in time* $\mathcal{O}(1.7088^n)$, *where n is the number of vertices of the input graph.*

Proof. Every maximal independent set of a graph G is also a dominating (not necessary minimum) set of G. First we compute any maximal independent set of G, which can be done by a greedy procedure in polynomial time. If the size of the maximal independent set found is larger than αn, for $0.2271 < \alpha < 0.22711$, by Theorem 3.11, we can compute $\gamma(G)$ in time $2^{n-\alpha n} \cdot n^{\mathcal{O}(1)} = \mathcal{O}(2^{0.7729n}) = \mathcal{O}(1.7088^n)$. If the size of the maximal independent set is at most αn, then we know that $\gamma(G) \leq \alpha n$, and by trying all subsets of size at most αn, we can find a minimum dominating set in time

$$\binom{n}{\alpha n} \cdot n^{\mathcal{O}(1)} = \mathcal{O}\left(\binom{n}{0.22711n}\right).$$

By making use of the formula for the entropy function (Lemma 3.13), we estimate

$$\mathcal{O}\left(\binom{n}{0.22711n}\right) = \mathcal{O}(2^{0.7729n}) = \mathcal{O}(1.7088^n).$$

$$\square$$

In Chap. 6 we present a faster algorithm for the MINIMUM DOMINATING SET problem based on the Branch & Reduce paradigm and Measure & Conquer analysis.

Exercise 3.15. Construct a greedy algorithm to compute some (no matter which) maximal independent set of an input graph $G = (V, E)$ in polynomial time. (Note that such a procedure is needed as part of the algorithm of Corollary 3.14).)

Exercise 3.16. The EXACT SET COVER problem is a covering problem. For a given universe $\mathcal{U}$ of elements and a collection $\mathcal{S}$ of (non-empty) subsets of $\mathcal{U}$, the task is to determine the minimum cardinality of a subset $\mathcal{S}' \subseteq \mathcal{S}$ which *partitions* $\mathcal{U}$, i.e.

$$\cup_{S \in \mathcal{S}'} S = \mathcal{U}$$

and for every $S, S' \in \mathcal{S}'$, if $S \neq S'$ then $S \cap S' = \emptyset$. Construct an exact algorithm solving the EXACT SET COVER problem on any input with n elements and m sets in time $\mathcal{O}(nm 2^n)$.

Exercise 3.17. The EXACT SAT problem (XSAT) is a variant of the SATISFIABILITY problem, where for a given CNF-formula, the task is to find a satisfying assignment such that in every clause exactly one literal is true. Show that XSAT can be solved in time $\mathcal{O}^*(2^m)$ on input CNF-formulas with m clauses.
Hint: Reduce XSAT to the EXACT HITTING SET problem by eliminating literals

of different signs. In EXACT HITTING SET we are given a universe $\mathcal{U}$ of elements and a collection $\mathcal{S}$ of (non-empty) subsets of $\mathcal{U}$, the task is to determine the minimum cardinality of a subset $\mathcal{U}' \subseteq \mathcal{U}$ such that $\mathcal{U}'$ hits every set $S \in \mathcal{S}$ exactly once, i.e. $|\mathcal{U}' \cap S| = 1$ for all $S \in \mathcal{S}$. Use the observation that EXACT HITTING SET and EXACT SET COVER are dual problems.

3.3 TSP on Graphs of Bounded Degree

In this section we revisit the dynamic programming algorithm for the TRAVELLING SALESMAN problem of Chap. 1.

To an input of TSP consisting of a set of cities $\{c_1, c_2, \ldots, c_n\}$ and a distance function d, we associate a graph G on vertex set $V = \{c_1, c_2, \ldots, c_n\}$. Two vertices c_i and c_j, $i \neq j$, are adjacent in G if and only if $d(c_i, c_j) < \infty$. We show that for graphs of bounded maximum degree, the dynamic programming algorithm for the travelling salesman problem of Chap. 1 runs in time $\mathcal{O}^*(c^n)$, for some $c < 2$. The proof is based on the observation that the running time of the dynamic programming algorithm is proportional to the number of connected vertex sets in a graph.

We call a vertex set $C \subseteq V$ of a graph $G = (V, E)$ *connected* if $G[C]$, the subgraph of G induced by C, is connected. While for a graph G on n vertices, the number of connected vertex sets in G can be as large as 2^n, for graphs of bounded degree it is possible to show that the maximum number of connected vertex subsets is significantly smaller than 2^n. The proof of this fact is based on the following lemma of Shearer.

Lemma 3.18 (Shearer's Lemma). *Let $\mathcal{U}$ be a finite set of elements with a collection $\mathcal{S} = \{S_1, S_2, \ldots, S_r\}$ of (non-empty) subsets of $\mathcal{U}$ such that every element $u \in \mathcal{U}$ is contained in at least δ subsets of $\mathcal{S}$. Let $\mathcal{T}$ be another family of subsets of $\mathcal{U}$. For each $1 \leq i \leq r$, we define the projection $\mathcal{T}_i = \{T \cap S_i : T \in \mathcal{T}\}$. Then*

$$|\mathcal{T}|^\delta \leq \prod_{i=1}^{r} |\mathcal{T}_i|.$$

For more information on Shearer's lemma we refer the reader to the Notes. For a proof we refer to Corollary 14.6.5 (p. 243) in the book of Alon and Spencer [4].

In the proof of the next lemma we use Jensen's inequality, a classical inequality on convex functions. For more information on Jensen's inequality the reader is referred to the Notes. Let I be an interval of $\mathbb{R}$. A real-valued function f defined on I is a *convex function* on I, if

$$f(\lambda x + (1 - \lambda)y) \leq \lambda f(x) + (1 - \lambda)f(y)$$

for all $x, y \in I$ and $0 \leq \lambda \leq 1$. Or equivalently, for any two points of the graph of f, the segment connecting these points, is above or on the graph of f.

Lemma 3.19 (Jensen's inequality). *Let f be a convex function on the interval I of* $\mathbb{R}^1$. *Then*

$$f(\sum_{i=1}^{n} \lambda_i x_i) \leq \sum_{i=1}^{n} \lambda_i f(x_i).$$

Using Shearer's lemma and Jensen's inequality we are able to upper bound the maximum number of connected vertex subsets in graphs of bounded maximum degree.

Lemma 3.20. *Let $G = (V,E)$ be a graph on n vertices of maximum degree Δ. Then the number of connected vertex subsets in G is at most*

$$(2^{\Delta+1} - 1)^{n/(\Delta+1)} + n.$$

Proof. The closed neighborhood of a vertex v, denoted by $N[v]$, consists of v and all neighbors of v: $N[v] = \{u \in V : \{u,v\} \in E\} \cup \{v\}$.

We start by defining sets S_v for all $v \in V$. Initially we assign $S_v := N[v]$. Then for every vertex v which is not of maximum degree, $d(v) < \Delta$, we add v to $\Delta - d(v)$ sets S_u which do not already contain v (it does not matter which sets S_u). Having completed this construction, every vertex v is contained in exactly $\Delta + 1$ sets S_u and

$$\sum_{v \in V} |S_v| = n(\Delta + 1).$$

Let $\mathcal{C}$ be the set of all connected vertex sets in G of size at least two. For every $v \in V$, the number of subsets in the projection $\mathcal{C}_v = \{C \cap S_v : C \in \mathcal{C}\}$ is at most $2^{|S_v|} - 1$. This is because for any connected set C of size at least two, $C \cap N[v] \neq \{v\}$, and thus the singleton set $\{v\}$ does not belong to $\mathcal{C}_v$. By Lemma 3.18,

$$|\mathcal{C}|^{(\Delta+1)} \leq \prod_{v \in V} (2^{|S_v|} - 1). \tag{3.5}$$

Let us define $f(x) = 2^x - 1$. It is quite easy to check that the function $-\log(f(x))$ is convex on the interval $[1, +\infty)$, and by Jensen's inequality, (Lemma 3.19) we have that

$$-\log f(\frac{1}{n} \sum_{i=1}^{n} x_i) \leq -\frac{1}{n} \log f(\sum_{i=1}^{n} x_i)$$

for any $x_i \geq 1$. In particular,

$$\frac{1}{n} \sum_{v \in V} \log f(|S_v|) \leq \log f(\frac{1}{n} \sum_{v \in V} |S_v|) = \log f(\Delta + 1).$$

By taking exponentials and combining with (3.5), we obtain

$$\prod_{v \in V} (2^{|S_v|} - 1) \leq (2^{\Delta+1} - 1)^n.$$

Therefore

$$|\mathcal{C}| \leq (2^{\Delta+1} - 1)^{n/(\Delta+1)}.$$

Finally, the number of connected sets of size one is at most n, which concludes the proof of the lemma. □

As soon as the bound on the maximum number of connected vertex subsets in a graph of bounded degree is established, the proof of the main result of this section becomes straightforward.

Theorem 3.21. *TSP on n-vertex graph of maximum degree Δ can be solved in time* $\mathcal{O}^*((2^{\Delta+1} - 1)^{n/(\Delta+1)})$.

Proof. We use the dynamic programming algorithm tsp provided in Chap. 1 (without any changes). The only observation is that if the set S in the recurrence (1.1) is a prefix of a tour for the salesman, then S is a connected vertex set of G. Thus the computation in (1.1) can be reduced to connected sets S.

Hence we may now modify the original TSP algorithm by running the dynamic programming algorithm tsp only over connected vertex sets S. Let us note that the connectivity test is easily done in polynomial (actually linear) time using depth first search.

By Lemma 3.20, the number of connected vertex subsets in a G of maximum degree Δ is at most $(2^{\Delta+1} - 1)^{n/(\Delta+1)}$, and the theorem follows. □

3.4 Partition into Sets of Bounded Cardinality

In Sect. 3.1.2, we discussed how to use dynamic programming to compute the chromatic number of a graph. This is a typical partition problem, i.e. partition of the vertex set of the graph into independent sets, and many such problems can be solved by the use of dynamic programming. In this section we consider the problem of counting k-partitions for sets of bounded cardinality, more precisely, each set $S \in \mathcal{S}$ of the input $(\mathcal{U}, \mathcal{S})$ has cardinality at most r, for some fixed integer $r \geq 2$.

Let $\mathcal{U}$ be a set of n elements and let $\mathcal{S}$ be a family of subsets of $\mathcal{U}$. Then a collection $S_1, S_2, \ldots, S_k$ of nonempty pairwise disjoint subsets is a *k-partition* of $(\mathcal{U}, \mathcal{S})$ if $S_i \in \mathcal{S}$ for all i with $1 \leq i \leq k$, and $S_1 \cup S_2 \cup \cdots \cup S_k = \mathcal{U}$. The task is to determine the number of unordered k-partitions of the input $(\mathcal{U}, \mathcal{S})$. (The number of ordered k-partitions of $(\mathcal{U}, \mathcal{S})$, i.e. the number of k-tuples $(S_1, S_2, \ldots, S_k)$ such that $\{S_1, S_2, \ldots, S_k\}$ is a k-partition of $(\mathcal{U}, \mathcal{S})$, is $OPT[\mathcal{U}, k] \cdot k!$.)

It is worth mentioning that sometimes the family $\mathcal{S}$ is given as a part of the input, but often $\mathcal{S}$ is given implicitly by a description which usually implies a polynomial (in $|\mathcal{U}|$) time algorithm to recognize the sets of $\mathcal{S}$. In this section we assume that there is a polynomial time algorithm verifying for each $S \subseteq \mathcal{U}$ whether $S \in \mathcal{S}$.

Counting problems are believed to be more difficult than their decision or optimization counterparts. A classical example is the problem of counting perfect matchings in a bipartite graph which is equivalent to the computation of the permanent of a matrix. While a perfect matching in a graph can be found in polynomial time, the counting version, even for bipartite graphs, is in #P, which is the counting counterpart of NP-hard problems.

A natural approach to count the k-partitions of an input $(\mathcal{U}, \mathcal{S})$ works as follows. Let $OPT[U; j]$ be the number of unordered j-partitions of $(U, \mathcal{S})$ for every nonempty subset $U \subseteq \mathcal{U}$ and $j \geq 1$. As in the previous sections, the value $OPT[\mathcal{U}, k]$, i.e. the number of unordered k-partitions of $(\mathcal{U}, \mathcal{S})$, can be computed by dynamic programming. Clearly, $OPT[U; 1] = 1$ if $U \in \mathcal{S}$, and $OPT[U; 1] = 0$ otherwise. For $j > 1$,

$$OPT[U; j] = \sum_{X \subseteq U, X \in \mathcal{S}} OPT[U \setminus X; j - 1] \qquad (3.6)$$

When $|\mathcal{S}|$ is small the running time required to compute (3.6) for all subsets of $\mathcal{U}$ is $\mathcal{O}^*(2^n |\mathcal{S}|)$. When $|\mathcal{S}|$ is large, the running time of the algorithm becomes $\mathcal{O}^*(3^n)$.

In Chaps 4 and 7, we present $\mathcal{O}^*(2^n)$ time algorithms counting the k-partitions of an input $(\mathcal{U}, \mathcal{S})$ by the use of inclusion-exclusion and subset convolution, respectively. No faster algorithm is known for counting k-partitions in general.

In this section we consider the problem of counting k-partitions when restricted to inputs $(\mathcal{U}, \mathcal{S})$ satisfying $|S| \leq r$ for every $S \in \mathcal{S}$, where $r \geq 2$ is a fixed integer. We show that it is possible to modify the dynamic programming algorithm above so as to count the k-partitions in time $\mathcal{O}^*((2 - \varepsilon)^n)$ for some $\varepsilon > 0$.

Theorem 3.22. *Let $\mathcal{U}$ be an n-element set, $r \geq 1$ be an integer, and $\mathcal{S}$ be a family of subsets of $\mathcal{U}$ such that every $S \in \mathcal{S}$ is of size at most r. Then for any $k \geq 1$, the number of k-partitions of $(\mathcal{U}, \mathcal{S})$ can be computed in time*

$$\mathcal{O}^*\left(\sum_{i=1}^{k} \sum_{j=i}^{ir} \binom{n-i}{j-i} \cdot |\mathcal{S}| \right).$$

Proof. We fix an ordering $u_1 < u_2 < \cdots < u_n$ of the elements of $\mathcal{U}$. This ordering establishes the lexicographical order $\prec$ over the subsets of $\mathcal{S}$. Thus for sets $S_1, S_2 \in \mathcal{S}$, we have $S_1 \prec S_2$ when either the minimum element of $S_1 \setminus S_2$ is less than the minimum element of $S_2 \setminus S_1$, or when $S_1 \subset S_2$. For example,

$$\{u_1, u_5\} \prec \{u_1, u_5, u_{10}\} \prec \{u_3, u_8\} \prec \{u_3, u_9\}.$$

We say that an *ordered k-partition* $(S_1, S_2, \ldots, S_k)$ is *lexicographically ordered* if $S_1 \prec S_2 \prec \cdots \prec S_k$. Let $\mathcal{L}_k$ be the set of all lexicographically ordered k-partitions of $(\mathcal{U}, \mathcal{S})$. The crucial observation is that $|\mathcal{L}_k|$ is equal to the number of (unordered) k-partitions of $(\mathcal{U}, \mathcal{S})$. Indeed, for every k-partition $\{S_1, S_2, \ldots, S_k\}$, the lexicographical ordering of sets S_i, $1 \leq i \leq k$, is unique. Also every lexicographically ordered k-partition $(S_1, S_2, \ldots, S_k)$ forms a unique unordered k-partition. Thus there is a bijection from $\mathcal{L}_k$ to the set of all unordered k-partitions, which means that the cardinalities of these sets are equal.

We compute $|\mathcal{L}_k|$ very similarly to (3.6). The main difference to (3.6) is that we do not have to go through all subsets of $\mathcal{U}$. We are looking only for lexicographically ordered sets, and this strongly reduces the search space and eliminates many non-relevant subsets. For example, for any $(S_1, S_2, \ldots, S_k) \in \mathcal{L}_k$, set S_1 must contain u_1. For $j > 1$, the set S_j must contain the minimum element not covered by $S_1, S_2, \ldots, S_{j-1}$.

Let us formally define over which subsets we proceed with the dynamic programming. The family of *relevant sets for* j, $\mathcal{R}_j$, $j \geq 1$, is defined recursively. We put

$$\mathcal{R}_1 = \{S \in \mathcal{S} : u_1 \in S\},$$

and for $j \geq 2$, the family $\mathcal{R}_j$ consists of all sets $S \cup S^*$ such that

- $S \cap S^* = \emptyset$;
- $S \in \mathcal{R}_{j-1}$;
- $S^* \in \mathcal{S}$;
- The minimum element of $\mathcal{U} \setminus \bigcup_{1 \leq i \leq j-1} S_i$ belongs to S_j.

Thus every set in $\mathcal{R}_j$ contains $u_1, \ldots, u_j$ and for every lexicographically ordered k-partition $(S_1, S_2, \ldots, S_k)$ of $\mathcal{U}$, the set $S_1 \cup \cdots \cup S_j$ is a relevant set for j.

For a relevant set $U \subseteq \mathcal{U}$ and $j \geq 1$, let $OPT[U; j]$ be the number of lexicographically ordered j-partitions of $(U, \mathcal{S})$. Thus $OPT[\mathcal{U}; k] = |\mathcal{L}_k|$. Then $OPT[U; 1] = 1$ if $U \in \mathcal{R}_1$, and $OPT[U; 1] = 0$ otherwise. For $j > 1$ and $U \in \mathcal{R}_j$,

$$OPT[U; j] = \sum_X OPT[U \setminus X; j-1], \tag{3.7}$$

where summation is taken over all sets $X \subseteq U$ such that

- $U \setminus X \in \mathcal{R}_{j-1}$;
- $X \in \mathcal{S}$;
- X contains the minimum element of $\mathcal{U} \setminus (U \setminus X)$.

Indeed, by the above discussions, every lexicographically ordered j-partition of a relevant set U is a union of $U' \cup X$, where U' is a relevant set for $j-1$, and $X \in \mathcal{S}$ is a set containing the minimum element of $\mathcal{U} \setminus U'$. Thus (3.7) correctly computes $OPT[U; j]$.

Consider the complexity of computing (3.7). We have to verify all subsets $X \subseteq U$ which belong to $\mathcal{S}$. For each such X, we check whether $U \setminus X \in \mathcal{R}_{j-1}$ and whether X contains the minimum element of $\mathcal{U} \setminus (U \setminus X)$. Thus the running time of the algorithm is

$$\mathcal{O}^*(|\mathcal{S}| \cdot (|\mathcal{R}_1| + |\mathcal{R}_2| + \cdots + |\mathcal{R}_k|)). \tag{3.8}$$

Finally, we claim that

$$|\mathcal{R}_i| \leq \sum_{j=i}^{ir} \binom{n-i}{j-i}. \tag{3.9}$$

Indeed, every set S from $\mathcal{R}_i$ contains $u_1, u_2, \ldots, u_i$. Also because the size of every set from $\mathcal{S}$ is at most r, we have that $|S| \leq ir$. Thus S contains at most $i(r-1)$ elements from $u_{i+1}, u_{i+2}, \ldots, u_n$ and (3.9) follows.

Putting together (3.8) and (3.9), we conclude the proof of the theorem. $\qquad \square$

To estimate the values given in Theorem 3.22, we use the classical combinatorial results on Fibonacci numbers. The sequence of *Fibonacci numbers* $\{F_n\}_{n\geq 0}$ is recursively defined as follows: $F(0) = 0$, $F(1) = 1$, and $F_{n+2} = F_{n+1} + F_n$. Thus, for example, $F(2) = 1$, $F(3) = 2$, $F(4) = 3$, $F(5) = 5$, etc. For more information on the Fibonacci numbers the reader is referred to the Notes. It is possible to give a closed form expression for Fibonacci numbers.

Lemma 3.23 (Fibonacci numbers). *For $n \geq 0$,*

$$F(n+1) = \left\lfloor \frac{\varphi^{n+1}}{\sqrt{5}} + \frac{1}{2} \right\rfloor,$$

where $\varphi = (1 + \sqrt{5})/2$ is the Golden Ratio.

We will need another identity for Fibonacci numbers.

Lemma 3.24.

$$F(n+1) = \sum_{k=1}^{\lfloor n/2 \rfloor} \binom{n-k}{k}.$$

Proof. Let a_n be the number of binary strings (i.e. sequences of 0s and 1s) of length n with no consecutive 0s. For example, below are all such strings of length 4

$$0111\ 0101\ 0110\ 1111$$
$$1011\ 1010\ 1101\ 1110$$

To prove the lemma, we will compute a_n in two different ways.

One can verify that $a_1 = 2$, $a_2 = 3$, $a_3 = 5$, $a_4 = 8$, and it is natural guess that $a_n = F_{n+2}$. Indeed, the number of strings of length n with no consecutive 0s which start from 1 is a_{n-1} and the number of such strings starting from 0 is a_{n-2}. Thus $a_n = a_{n-1} + a_{n-2}$, and we have that $a_n = F_{n+2}$.

On the other hand, the number of strings of length $k+r$ with k ones and r zeros such that no two zeros are consecutive, is equal to

$$\binom{k+1}{r}.$$

Indeed, every such string is formed from a string of length k of 1s by inserting r 0s. But there are exactly $k+1$ places to insert 0s: before the first 1, after the last 1, or between adjacent 1s. Thus the number of such strings of length n with exactly r zeros is

$$\binom{n-r+1}{r},$$

and we conclude that

$$a_n = \sum_{r=0}^{n} \binom{n-r+1}{r}.$$

Putting together the two identities for a_n, we have that

$$a_n = F_{n+2} = \sum_{r=0}^{n} \binom{n-r+1}{r}$$

and by replacing n by $n-1$, we have that

$$F_{n+1} = \sum_{r=0}^{n} \binom{n-r}{r}.$$

Finally, for $r > \lfloor n/2 \rfloor$, every string of length n with r 0s has two consecutive 0s, and thus

$$F(n+1) = \sum_{k=1}^{\lfloor n/2 \rfloor} \binom{n-k}{k}.$$

$\square$

Let us give an application of Theorem 3.22. Counting perfect matchings in an n-vertex graph is the case of Theorem 3.22 with $r = 2$ and $k = n/2$ (we assume that n is even).

Corollary 3.25. *The number of perfect matchings in an n-vertex graph can be computed in time* $\mathcal{O}(1.6181^n)$.

Proof. The running time of the algorithm is proportional, up to a polynomial factor, to

$$\max_{1 \leq i \leq n/2} \sum_{j=i}^{ir} \binom{n-i}{j-i} = \max_{1 \leq i \leq n/2} \sum_{j=i}^{2i} \binom{n-i}{j-i}.$$

If $i \geq n/3$, then

$$\sum_{j=i}^{2i} \binom{n-i}{j-i} \leq 2^{n-i} \leq 2^{2n/3}.$$

If $i \leq n/3$, then

$$\sum_{j=i}^{2i} \binom{n-i}{j-i} = \sum_{k=1}^{i} \binom{n-i}{k} \leq i \cdot \binom{n-i}{i}.$$

By Lemma 3.24,

$$\sum_{k=1}^{\lfloor n/2 \rfloor} \binom{n-k}{k} = F(n+1),$$

where

$$F(n+1) = \left\lfloor \frac{\varphi^{n+1}}{\sqrt{5}} + \frac{1}{2} \right\rfloor$$

is the $(n+1)$th Fibonacci number and $\varphi = (1+\sqrt{5})/2$ is the Golden Ratio. Then

r	2	3	4	5	6	7	20	50
c_r	1.6181	1.7549	1.8192	1.8567	1.8813	1.8987	1.9651	1.9861

Fig. 3.2 Running time $\mathcal{O}(c_r{}^n)$ for some values of r

$$\sum_{j=i}^{2i} \binom{n-i}{j-i} \leq i \cdot \binom{n-i}{i}$$

$$< i\varphi^{n+1} = \mathcal{O}(n \cdot 1.6181^n).$$

□

In Fig. 3.2 a table with the bases of the exponential running time for certain values of r is given.

Notes

Dynamic programming is one of the fundamental techniques for designing algorithms. The term dynamic programming is due to Richard Bellman [15]. The textbooks of Cormen, Leiserson, Rivest and Stein [52] and Kleinberg and Tardos [129] provide nice introductions to dynamic programming.

One of the earliest exact algorithms for a permutation problem are dynamic programming algorithms of Bellman [16, 16] and of Held and Karp [111] for the TRAVELLING SALESMAN (TSP) problem. The dynamic programming algorithm for the DIRECTED FEEDBACK ARC SET problem is a classical result from 1964 due to Lawler [149].

Dynamic programming over subsets does not seem to work for every permutation problem. For example, it is not clear how to apply it directly to the BANDWIDTH MINIMIZATION problem (see also Sect. 11.1) or to the SUBGRAPH ISOMORPHISM problem (see also Sect. 4.4). It is an open problem whether the BANDWIDTH MINIMIZATION problem can be solved in time $\mathcal{O}^*(2^n)$. For SUBGRAPH ISOMORPHISM no algorithm of running time $2^{o(n \log n)}$ is known. Improving the $\mathcal{O}^*(2^n)$ running time for solving the HAMILTONIAN PATH problem and the TSP problem is a long standing open problem.

For some other problems, like TREEWIDTH and PATHWIDTH (which can be seen as permutation problems) faster algorithms than the $\mathcal{O}^*(2^n)$ dynamic programming algorithms are known. The treewidth of a graph can be computed in time $\mathcal{O}(1.7549^n)$ [90, 95]. Suchan and Villanger showed how to compute the pathwidth of a graph in time $\mathcal{O}(1.9657^n)$ [211]. An attempt to classify permutation problems which can be solved by dynamic programming over subsets together with some implementations can be found in [31].

For a nice treatment of computational and combinatorial issues of Stirling formula and Fibonacci numbers we recommend the book of Graham, Knuth, and Patashnik [104]. Theorem 3.10 is from [91]. Theorem 3.11 and Corollary 3.14 are due to Liedloff [152]. In Chap. 6, a faster algorithm for MINIMUM DOMINATING SET of running time $\mathcal{O}(1.5259^n)$ is provided by making use of Measure & Conquer.

The first dynamic programming algorithm computing an optimal graph coloring of running time $\mathcal{O}(2.4423^n)$ was published by Lawler in 1976 [150]. The running time of Lawler's algorithm was improved for the first time only in 2001 by Eppstein [69, 70] and then by Byskov [42] who refined Eppstein's idea and obtained a running time of $\mathcal{O}(2.4023^n)$. All these results were significantly improved by Björklund, Husfeldt and Koivisto by making use of inclusion-exclusion (see Chap. 4) [24, 30, 137]. Deciding whether a graph is 3-colorable can be done by making use of the algorithm of Beigel and Eppstein [14] in time $\mathcal{O}(1.3289^n)$. Exact algorithms for counting 3- and 4-colorings can be found in [6, 80, 99].

Jensen's inequality can be found in any classical book on analysis, see, e.g. [193]. Shearer's Lemma appeared in [49]. For a proof, see the book of Alon and Spencer [4]. The lemma was used by Björklund, Husfeldt, Kaski and Koivisto [28] to obtain a $\mathcal{O}^*((2-\varepsilon)^n)$ algorithms for coloring and other partition and covering problems on graphs of bounded degrees. The proof of Theorem 3.21 is taken from [28]. The running time in Theorem 3.21 can be improved by considering not only connected sets, but also dominating sets with specific properties. For graphs of maximum degree 3, improved algorithms for TSP were obtained by Eppstein [73] and Iwama [123]. See also the work of Björklund, Husfeldt, Kaski and Koivisto [29] for $\mathcal{O}^*((2-\varepsilon)^n)$ time algorithms for different partition problems on graphs with bounded vertex degrees. Another class of graphs on which deciding Hamiltonicity can be done faster than $\Theta(2^n)$ is the class of claw-free graphs [40]. Theorem 3.22 and Corollary 3.25 are due to Koivisto [138]. For a version of EXACT SET COVER, where every set is of size at most k, Björklund obtained a randomized algorithm of running time $\mathcal{O}(c_k^n)$, where $c_3 \leq 1.496$, $c_4 \leq 1.642$, and $c_5 \leq 1.721$ [23].

Chapter 4
Inclusion-Exclusion

Inclusion-exclusion is a fascinating technique used in the design of fast exponential time algorithms. It is based on the inclusion-exclusion principle which is a fundamental counting principle in combinatorics; it allows us to count combinatorial objects in a somewhat indirect way that is applied in particular when direct counting is not possible. This counting principle is the main tool when designing inclusion-exclusion algorithms. It seems that this algorithm design paradigm is suited very well to constructing fast exponential time algorithms since it naturally produces exponential time algorithms.

Similar to dynamic programming, inclusion-exclusion based algorithms go through all possible subsets but the significant difference is that they do not require exponential space. Thus like dynamic programming, inclusion-exclusion algorithms can often be used not only to solve NP-hard problems but also to solve seemingly harder counting problems. Combined with dynamic programming, inclusion-exclusion is the basis of several fast transforms, i.e. transformations of a set function to another set function. In this chapter we use such a combination to explain the breakthrough $\mathcal{O}^*(2^n)$ graph coloring algorithms of Björklund and Husfeldt [24] and Koivisto [137]. In Chap. 7 we give several fast transforms based on inclusion-exclusion.

4.1 The Inclusion-Exclusion Principle

In this section we present the inclusion-exclusion principle of counting. We start with a simple example. Let S be a set and A, B be some properties of elements of S. Each of the elements of S can have one, both or none of the properties. We denote by $N(0)$ the number of elements having no properties, by N_A (N_B) the number of elements with property A (B), and by $N_{A \cap B}$ the number of elements with both properties. Because each of the elements with both properties is counted twice in the sum $N_A + N_B$, we have that

$$N(0) = N - (N_A + N_B) + N_{A \cap B}.$$

F.V. Fomin, D. Kratsch, *Exact Exponential Algorithms*, Texts in Theoretical Computer Science. An EATCS Series, DOI 10.1007/978-3-642-16533-7_4, © Springer-Verlag Berlin Heidelberg 2010

The following theorem generalizes our example.

Theorem 4.1. *Given a collection of N combinatorial objects and properties* $P(1)$, $P(2),\ldots,P(n)$ *such that for every* $i \in \{1,2,\ldots,n\}$ *each object does or does not have property* $P(i)$. *Hence each of the objects has some subset of those properties: none, several or all of* $P(1),P(2),\ldots,P(n)$.

For every $r \geq 1$ *and* $\{i_1,i_2,\ldots,i_r\} \subseteq \{1,2,\ldots,n\}$, *we denote by* $N_{i_1,i_2,\ldots,i_r}$ *the number of objects having (at least) all the properties* $P(i_1),P(i_2),\ldots$, *and* $P(i_r)$. *Then* $N(0)$, *the number of objects having none of the properties, can be determined by the following formula of inclusion-exclusion:*

$$N(0) = N - \sum_{i=1}^{n} N_i + \sum_{i_1 < i_2} N_{i_1,i_2} - \sum_{i_1 < i_2 < i_3} N_{i_1,i_2,i_3} + \cdots$$

$$\cdots + (-1)^j \sum_{i_1 < i_2 < \cdots < i_j} N_{i_1,i_2,\ldots,i_j} + \cdots + (-1)^n N_{1,2,\ldots,n} \qquad (4.1)$$

Proof. For every object we verify how often it is counted on each of the two sides of equation (4.1).

If an object has none of the properties then it is counted once on the left hand side and once on the right hand side (for N). Now suppose that an object has precisely the properties $P(j_1),P(j_2),\ldots,P(j_s)$, $s \geq 1$. On the right hand side such an object is counted precisely for those $N_{i_1,i_2,\ldots,i_r}$ satisfying $\{j_1,j_2,\ldots,j_s\} \subseteq \{i_1,i_2,\ldots,i_r\}$, i.e. it is counted once for $N_{i_1,i_2,\ldots,i_r}$ for each superset $\{i_1,i_2,\ldots,i_r\}$ of $\{j_1,j_2,\ldots,j_s\}$. Hence for each $r \geq s$, such an object is counted $\binom{n-s}{r-s}$ times, since s objects, i.e. $j_1,j_2,\ldots,j_s$ of all sets to count, are fixed in advance. Finally, due to the alternation of the signs $+$ and $-$ in equation (4.1), the contribution of such an object is

$$\sum_{\{j_1,j_2,\ldots,j_s\} \subseteq \{i_1,i_2,\ldots,i_r\}} N_{i_1,i_2,\ldots,i_r} = \sum_{r=s}^{n} (-1)^r \binom{n-s}{r-s}$$

$$= \sum_{k=0}^{n-s} (-1)^k \binom{n-s}{k} = 0.$$

This completes the proof.   □

In combinatorics the inclusion-exclusion principle is used when it is difficult to determine the number of objects without any of the properties directly while it is much easier to determine all the numbers $N_{i_1,i_2,\ldots,i_r}$. While inclusion-exclusion is a powerful and relatively simple counting principle, algorithms using inclusion-exclusion have to run through all subsets of the ground set, which results in exponential running time.

The following theorem is another version of the inclusion-exclusion principle which is somewhat complementary to the one given above in Theorem 4.1.

Theorem 4.2. *Given a collection of N combinatorial objects and properties* $Q(1)$, $Q(2),\ldots,Q(n)$ *such that each of the objects has a subset of those properties. For* *any subset $W \subseteq \{1,2,\ldots,n\}$, let $N(W)$ be the number of objects having none of the properties $Q(w)$ with $w \in W$ (but possibly some of the others). Let X be the number of objects having all properties $Q(1), Q(2), \ldots, Q(n)$. Then*

$$X = \sum_{W \subseteq \{1,2,\ldots,n\}} (-1)^{|W|} N(W) \tag{4.2}$$

Proof. To apply Theorem 4.1, we say that an object has property $P(i)$ if and only if it does not have property $Q(i)$, where $1 \le i \le n$. Hence the number of objects with properties $P(i_1), P(i_2), \ldots, P(i_r)$ (and maybe some other properties) is equal to the number of objects having none of the properties $Q(i_1), Q(i_2), \ldots, Q(i_r)$. Then Theorem 4.1 implies (4.2). □

Let us emphasize that when inclusion-exclusion is used to construct fast exponential time algorithms usually the version in Theorem 4.2 is applied. How to choose objects and properties for a particular (counting) problem will become clear in the subsequent sections.

4.2 Some Inclusion-Exclusion Algorithms

In this section we give several examples of how inclusion-exclusion ideas can be used to design exact algorithms. The idea of the first example, computing the permanent of the matrix or counting perfect matchings in a bipartite graph, goes back to the work of Ryser [194]. The second example, the HAMILTONIAN PATH problem, is a special case of the TSP problem, which was discussed in Chap. 1 and 3. The application of inclusion-exclusion to TSP and its variants like the Hamiltonian cycle and the Hamiltonian path problem was rediscovered several times [136, 127, 11]. Our third example is the BIN PACKING problem.

The running time of the inclusion-exclusion algorithm is almost the same as the time used by the dynamic programming algorithm similar to the one for TSP from Chap. 1. However the dynamic programming algorithm needs exponential space while the inclusion-exclusion algorithm uses polynomial space. We will discuss the importance of space bounds and the possibilities of time and space exchange in Chap. 10.

4.2.1 Computing the Permanent of a Matrix

Let $A = [a_{ij}]$, $i, j \in \{1, \ldots, n\}$, be a binary $n \times n$ matrix, i.e. a matrix in which each entry a_{ij} is either zero or one. An important characteristic of a matrix is its *permanent* defined as

$$\text{perm}(A) = \sum_{\pi \in S_n} \prod_{i=1}^{n} a_{i,\pi(i)},$$

where S_n is the set of all permutations of $\{1,\ldots,n\}$. For example, the permanent of matrix

$$A = \begin{pmatrix} a_{11} & a_{12} & a_{13} \\ a_{21} & a_{22} & a_{23} \\ a_{31} & a_{32} & a_{33} \end{pmatrix}$$

is

$$\mathrm{perm}(A) = a_{11}a_{22}a_{33} + a_{11}a_{32}a_{23} + a_{21}a_{12}a_{33} + a_{21}a_{32}a_{13}$$
$$+ a_{31}a_{12}a_{23} + a_{31}a_{22}a_{13}.$$

While the formula of the permanent looks very similar to the formula of the determinant, the computational complexity of these two characteristics is drastically different. The trivial algorithm computing the permanent of a matrix would be to try all possible permutations π, and compute for each permutation the corresponding sum, which results in running time $\mathcal{O}^*(n!)$.

Computing the permanent of a binary matrix can be expressed in the language of Graph Theory as a problem of counting perfect matchings in bipartite graphs. A matching M of a graph $G = (V,E)$ is *perfect* if M is an edge cover of G. In other words, every vertex of G is incident to precisely one edge of M. Deciding whether a graph has a perfect matching is a classical combinatorial problem and is solvable in polynomial time. However, counting perfect matchings is known to be a #P-complete problem.

Let $G = (V,E)$ be a bipartite graph with bipartition (X,Y), $X = \{x_1,x_2,\ldots,x_n\}$, $Y = \{y_1,y_2,\ldots,y_n\}$. Let $A_G = [a_{ij}]$, $i, j \in \{1,\ldots,n\}$ be the adjacency matrix of G, the $n \times n$ binary matrix whose entry a_{ij} is 1 if $\{x_i,y_j\} \in E$ and 0 otherwise. Let us define the characteristic function

$$f(x_i,y_j) = \begin{cases} 1 & \text{if } \{x_i,y_j\} \in E, \\ 0 & \text{otherwise.} \end{cases}$$

Then the number of perfect matchings in G is equal to

$$\sum_{\pi \in S_n} \prod_{i=1}^{n} f(x_i, y_{\pi(i)}).$$

The main observation here is that for every permutation $\pi \in S_n$, a set of edges $M = \{\{x_1,y_{\pi(1)}\}, \{x_2,y_{\pi(2)}\},\ldots, \{x_n,y_{\pi(n)}\}\}$ is a perfect matching if and only if $a_{1\pi(1)}a_{2\pi(2)} \cdots a_{n\pi(n)} = 1$. Thus the number of perfect matchings in G is equal to

$$\sum_{\pi \in S_n} \prod_{i=1}^{n} a_{i\pi(i)} = \mathrm{perm}(A^G).$$

An example is given in Fig. 4.1.

Using inclusion-exclusion and the scenario of Theorem 4.2 we prove the following lemma.

$$A^G = \begin{pmatrix} 1 & 0 & 0 \\ 1 & 1 & 1 \\ 0 & 1 & 1 \end{pmatrix}$$

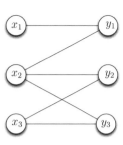

Fig. 4.1 A bipartite graph G and its adjacency matrix. The graph G has two perfect matchings, namely, $\{x_1,y_1\},\{x_2,y_2\},\{x_3,y_3\}$ and $\{x_1,y_1\},\{x_2,y_3\},\{x_3,y_2\}$. The permanent of A^G is 2.

Lemma 4.3. *Let G be a bipartite graph and A^G be its adjacency matrix. Then*

$$\operatorname{perm}(A^G) = \sum_{W \subseteq \{1,2,\dots,n\}} (-1)^{|W|} \prod_{i=1}^{n} \sum_{j \notin W} f(i,j).$$

Proof. Following the notation of Theorem 4.2, we define the collection N of the following objects. Each object $M \in N$ is a set of n edges of G, such that for every $i \in \{1,2,\dots,n\}$, the vertex x_i is an endpoint of some edge of M. The property $Q(j)$ is that the vertex y_j is an endpoint of some edge of M. In this language, an object M has all properties $Q(1),Q(2),\dots,Q(n)$ iff it is a perfect matching of G. For $W \subseteq \{1,2,\dots,n\}$, the number of objects $N(W)$ having none of the properties $Q(w)$, $w \in W$, is equal to

$$N(W) = \prod_{i=1}^{n} \sum_{j \notin W} f(i,j).$$

Indeed, for every vertex x_i, to form an object M, we can select exactly one edge with an endpoint x_i whose second endpoint is not in W. Now, by (4.2), the number X of objects having all properties, i.e. the number of perfect matchings in G or the permanent of A^G, is equal to

$$\operatorname{perm}(A^G) = X =$$

$$\sum_{W \subseteq \{1,2,\dots,n\}} (-1)^{|W|} N(W) = \sum_{W \subseteq \{1,2,\dots,n\}} (-1)^{|W|} \prod_{i=1}^{n} \sum_{j \notin W} f(i,j).$$

$\square$

Putting all together, we obtain the following theorem.

Theorem 4.4. *The permanent of a binary $n \times n$ matrix can be computed in time $\mathcal{O}^*(2^n)$ and polynomial space.*

Proof. For a binary matrix A, we construct the bipartite graph G, such that A is its adjacency matrix. Then by Lemma 4.3,

$$\mathrm{perm}(A^G) = \sum_{W \subseteq \{1,2,\ldots,n\}} (-1)^{|W|} \prod_{i=1}^{n} \sum_{j \notin W} f(i,j). \qquad (4.3)$$

We compute the permanent by making use of this formula. Each of the values

$$\prod_{i=1}^{n} \sum_{j \notin W} f(i,j)$$

is computable in polynomial time, and thus the running time required to compute the permanent is $\mathcal{O}^*(2^n)$. $\qquad \square$

4.2.2 Directed Hamiltonian Path

In the DIRECTED HAMILTONIAN s,t-PATH problem we are given a directed and simple graph $G = (V,E)$ with vertex set $\{s,t\} \cup \{v_1, v_2, \ldots, v_n\}$ and edge set E, $|E| = m$. The task is to determine whether there is a Hamiltonian path from s to t in G, i.e. a directed path $P = (s,\ldots,t)$ of length $n+1$ in G such that each vertex of G occurs precisely once in P.

To solve the decision problem DIRECTED HAMILTONIAN s,t-PATH we actually solve the corresponding counting problem.

Theorem 4.5. *The number of directed Hamiltonian s,t-paths in a graph can be computed in time $\mathcal{O}^*(2^n)$ and in polynomial space.*

Proof. Our algorithm will determine the number of Hamiltonian paths from s to t in G. To apply the inclusion-exclusion principle of counting, we use the scenario of Theorem 4.2. The objects are directed walks from s to t of length $n+1$. A *directed walk* from s to t of length $n+1$ is a sequence $s, v_1, \ldots, v_n, t$ of vertices such that every pair of consecutive vertices is connected by an edge; vertices and edges may show up repeatedly in a walk. A walk (or an object) has property $Q(i)$ if it contains vertex v_i. Hence for every subset $W \subseteq \{1,2,\ldots,n\}$, $N(W)$ is the number of objects having none of the properties $Q(i)$, for all $i \in W$. In other words, $N(W)$ is the number of directed walks of length $n+1$ from s to t containing no vertex v_i with $i \in W$. We define X to be the number of directed walks of length $n+1$ containing all vertices of $\{v_1, v_2, \ldots, v_n\}$. Thus X is the number of Hamiltonian paths from s to t. By the inclusion-exclusion principle as given in Theorem 4.2, we have that

$$X = \sum_{W \subseteq \{1,2,\ldots,n\}} (-1)^{|W|} N(W). \qquad (4.4)$$

Now all that remains is to compute $N(W)$ for every subset $W \subseteq \{1,2,\ldots,n\}$. For every fixed subset W this is easy, and actually can be performed in polynomial time. There are several ways of computing $N(W)$.

One way is to use adjacency matrices. Let A be the adjacency matrix of $G \setminus W$. Recall that in the kth power A^k of A, the entry at the intersection of row i and column j counts the number of walks with $k+1$ vertices in $G \setminus W$ that start at vertex i and end at vertex j. Therefore, the number $N(W)$ can be extracted from A^{n+1} which can be computed in polynomial time.

Another way of computing $N(W)$ is via dynamic programming. For a subset $W \subseteq \{1, 2, \ldots, n\}$ and vertex $u \in \{v_i : i \notin W\} \cup \{t\}$, we define $P_W(u, k)$ to be the number of directed walks of length k from s to u avoiding all vertices v_i with $i \in W$. With such a notation we have $N(W) = P_W(t, n+1)$.

The dynamic programming algorithm is based on the following recurrence.

$$P_W(u, 0) = \begin{cases} 0, & \text{if } u \neq s \\ 1, & \text{if } u = s \end{cases}$$

$$P_W(u, k) = \sum_{\{v_i : (v_i, u) \in E, \text{ and } i \notin W\}} P_W(v_i, k-1), \text{ for } 1 \leq k \leq n+1.$$

For each subset W, the dynamic programming algorithm based on the above recurrence performs $\mathcal{O}(nm)$ operations with integer-valued variables of the type $P_W(u, k)$. Thus the running time on the unit-cost RAM is $\mathcal{O}(nm)$. On the log-cost RAM the running time of this step is slightly different: The number of walks is at most 2^m, and thus each integer-value of a variable $P_W(u, k)$ can be encoded by making use of $\mathcal{O}(m)$ bits resulting in $\mathcal{O}(nm^2)$ time.

Note that at a typical point during the execution of the algorithm it stores for all suitable vertices u, the values $P_W(u, k)$ to be computed and the values $P_W(u, k-1)$ needed for the computation of the values of $P_W(u, k)$. Hence the required space is $\mathcal{O}(nm)$.

The overall inclusion-exclusion algorithm needs to plug all the 2^n values of $N(W)$ computed by the polynomial time dynamic programming algorithm into the formula (4.4). Hence its running time is $\mathcal{O}(nm2^n)$. Moreover the algorithm uses only space $\mathcal{O}(nm)$ because it may use the same space for each computation of $N(W)$, $(-1)^{|S|}N(S)$ and the corresponding partial sum. $\qquad \square$

Self-reduction. It is natural to wonder whether such an algorithm counting the Hamiltonian paths from s to t via an inclusion-exclusion formula and via counting s, t-walks can actually compute at least one Hamiltonian path from s to t if there is one. Indeed using an inclusion-exclusion algorithm to compute a (optimal) solution can often be done by a method called *self-reduction*. For the Hamiltonian path it can be done at the cost of an additional multiplicative factor $\log n$ and still within polynomial space.

Let us sketch the idea. By Theorem 4.5, we can count the number of Hamiltonian paths in time $\mathcal{O}(p(n)2^n)$, where $p(n)$ is some polynomial in n. If the algorithm outputs that the number of Hamiltonian paths in some input directed graph G is positive, we know that there exists at least one such path in G. We take an edge e of the input graph G, delete it from G and run the counting algorithm on the modified instance. If the number of paths in the new graph is positive, then we conclude that there is at least one Hamiltonian path in G that does not contain e. In this case we proceed recursively, i.e. pick another edge of the new graph, remove it, count the number of paths in the reduced instance, etc. If the graph $G \setminus \{e\}$ has no Hamiltonian path, we conclude that every Hamiltonian path in G goes through e. In this case we put e back into G, label it as an edge of the Hamiltonian path, so we never try to remove it again from the graph, and recursively try to remove another edge. By the end of this procedure we are left with a Hamiltonian path of G. However, such an approach requires to run the counting algorithm m times, resulting in running time $\mathcal{O}(m \cdot p(n) \cdot 2^n)$.

We first show how to decrease the number of runs from m to n. Instead of trying all edges, we start from trying only edges going from s. There are at most $n - 1$ such edges, and if for some of the edges, say (s, s'), we have that the amounts of Hamiltonian paths from s to t in graphs G and $G \setminus \{(s, s')\}$ are different, we know that there is at least one path in G using (s, s'). Moreover, in this path (s, s') is the only edge going from or to s. Thus in this case we can remove s with all incident edges and proceed constructing a Hamiltonian path from s' to t in $G \setminus \{s\}$ recursively. For a subgraph F of G on k vertices, we count the number of paths in F in time $\mathcal{O}(p(k) \cdot 2^k) = \mathcal{O}(p(n) \cdot 2^k)$, and thus the running time of the algorithm can be estimated as

$$\mathcal{O}(n \cdot p(n) \cdot (2^n + 2^{n-1} + \cdots + 2^0)) = \mathcal{O}(n \cdot p(n) \cdot 2^n).$$

Finally, to speed up the algorithm, we can search for an edge (s, s') whose removal changes the number of paths by performing binary search. We partition the set of edges incident to s into sets A and B of equal sizes and run counting algorithm on the graph $G \setminus A$ resulting from G by removing all edges from A. If the number of paths in G and $G \setminus A$ are the same, then (s, s') must be in B. Otherwise, it should be in A. In this way, we can find (s, s') in $\mathcal{O}(\log n \cdot p(n) \cdot 2^n)$ time, resulting in the running time

$$\mathcal{O}(\log n \cdot p(n) \cdot (2^n + 2^{n-1} + \cdots + 2^0)) = \mathcal{O}(\log n \cdot p(n) \cdot 2^n).$$

We would like to emphasize that with few modifications the algorithm to count the s, t-Hamiltonian paths of a directed graph also works for the Hamiltonian cycle problem on directed and also on undirected graphs. It also can easily be modified to solve the TSP problem. However when distances between cities might be exponential in n (the number of cities), then the algorithm will require exponential space as well. More precisely, with the maximum distance between cities W, the running time of the inclusion-exlcusion algorithm is $\mathcal{O}^*(2^n W)$, while the space required is $\mathcal{O}^*(W)$. Since the input length is $\mathcal{O}(n \log W)$, the space requirement is exponential. The existence of a polynomial space algorithm solving the TSP problem in time $\mathcal{O}^*(2^n)$ is an interesting open problem. In Chap. 10 we discuss techniques allowing us to solve the TSP problem in polynomial space, but with worse running time.

4.2.3 Bin Packing

Here is another example of how to apply inclusion-exclusion to solve a decision problem.

Bin Packing. In the BIN PACKING problem, we are given a positive integer bin capacity B, a positive integer k giving the number of available bins, and n items, where the size of item i is given by a positive integer $s(i)$. The task is to determine whether the set of items can be partitioned into sets $U_1, U_2, \ldots, U_k$ such that the sum of the sizes of the items in each U_j, $1 \leq j \leq k$, is at most B.

Theorem 4.6. *There is an algorithm that decides in time $\mathcal{O}(nB2^n)$ and space $\mathcal{O}(\max_{i=1}^n s(i))$ whether the bin packing problem with n items has a solution.*

Proof. A partition $U_1, U_2, \ldots, U_k$ of the items is called a *feasible solution* if the sum of the sizes of the items in each U_j is at most B. Since only the decision problem needs to be solved, we may relax the notion of a feasible solution such that items may appear more than once either in the same bin or in different bins. Thus if there is a relaxed feasible solution then there is also a feasible solution, and therefore it is sufficient to decide whether there is a relaxed feasible solution, and this will be done by counting the relaxed feasible solutions.

A relaxed feasible solution may be viewed as an ordered set of k finite lists of elements from $\{1, 2, \ldots, n\}$ such that

(a) for each of such list $a_1, a_2, \ldots, a_p$, $\sum_{h=1}^p s(a_h) \leq B$, and
(b) each of the elements of $\{1, 2, \ldots, n\}$ appears in at least one of the list.

To count the number of relaxed feasible solutions, we apply the inclusion-exclusion principle of Theorem 4.2. To apply Theorem 4.2, we define objects as ordered sets of k finite lists of elements from $\{1, 2, \ldots, n\}$ such that if $a_1, a_2, \ldots, a_p$ is one of the lists, then $\sum_{h=1}^p s(a_h) \leq B$. We specify the properties $Q(1), Q(2), \ldots, Q(n)$ of objects as follows. For $w \in \{1, 2, \ldots, n\}$, an object has the property $Q(w)$ if at least one of its lists contains the element w. For a subset $W \subseteq \{1, 2, \ldots, n\}$, we define $N(W)$ to be the number of objects not having property $Q(w)$, for all $w \in W$. Then by Theorem 4.2, the number of all relaxed feasible solutions, which is the number of objects possessing all properties, is

$$X = \sum_{W \subseteq \{1,2,\ldots,n\}} (-1)^{|W|} N(W).$$

For $W \subseteq \{1, 2, \ldots, n\}$, we define $A(W)$ as the number of lists $a_1, a_2, \ldots, a_p$ of elements not in W such that $\sum_{h=1}^p s(a_h) \leq B$. Then $N(W) = A(W)^k$ because the list for each of the k bins can contain only items for which the sum of their sizes is at most B. We denote by $P_W(j)$, $0 \leq j \leq B$, the number of lists $a_1, a_2, \ldots, a_p$ of elements not in W such that $\sum_{h=1}^p s(a_h) = j$. Hence $A(W) = \sum_{j=0}^B P_W(j)$.

It is convenient to extend the definition of $P_W(\ell)$ for negative ℓ by putting

$$P_W(\ell) = 0 \text{ for all } \ell < 0.$$

For every fixed W, there is only one (empty) list with sums of its elements equal to 0. Thus

$$P_W(0) = 1.$$

The values $P_W(j)$, $j \geq 1$, can be computed by a simple dynamic programming algorithm based on the following recurrence

$$P_W(j) = \sum_{i \notin W} P_W(j - s(i)).$$

Hence, for every fixed W, $N(W)$ can be computed in time $\mathcal{O}(nB)$. The space required for the computation is $\mathcal{O}(\max_{i=1}^n s(i))$ because only the last $\max_{i=1}^n s(i)$ values of $P_W(i)$ have to be stored for computing $P_W(j)$. Consequently the number of relaxed feasible solutions can be computed using time $\mathcal{O}(nB2^n)$ and space $\mathcal{O}(\max_{i=1}^n s(i))$.   □

4.3 Coverings and Partitions

The BIN PACKING problem can be seen as a covering problem, where one wants to cover the set $\{1, 2, \ldots, n\}$ with at most k sequences such that the sum of the elements in every sequence is at most B. While for the BIN PACKING problem the inclusion-exclusion algorithm requires the same time as known dynamic programming algorithms, for some of the covering problems discussed below inclusion-exclusion allows a significant speed-up.

In this section we present inclusion-exclusion algorithms for covering and partition problems. Such an algorithm is typically significantly faster than the best known dynamic programming algorithm for the corresponding problem. In particular, we present a $\mathcal{O}^*(2^n)$ time algorithm to compute the chromatic number of a graph and to solve the well-known coloring problem. These results were obtained independently in 2006 by Björklund and Husfeldt [24] and Koivisto [137].

Let us recall (see page 36) that in the MINIMUM SET COVER problem (MSC) we are given a universe $\mathcal{U}$ of elements and a collection $\mathcal{S}$ of (non-empty) subsets of $\mathcal{U}$, and the task is to find a subset of $\mathcal{S}$ of minimum cardinality covering all elements of $\mathcal{U}$.

Let $\mathcal{U}$ be a set of n elements and let $\mathcal{S}$ be a family of subsets of $\mathcal{U}$. We shall often denote such an input by $(\mathcal{U}, \mathcal{S})$. Additionally we assume that (all the elements of) $\mathcal{S}$ can be enumerated in time $\mathcal{O}^*(2^n)$. Typically $\mathcal{S}$ is implicitly defined by a polynomial time computable predicate. This additional assumption is needed to guarantee that the overall running time $\mathcal{O}^*(2^n)$ of the inclusion-exclusion algorithm can be established.

We often refer to a set cover of cardinality k as to a k-cover. In other words, $S_1, S_2, \ldots, S_k$ is a k-cover of $(\mathcal{U}, \mathcal{S})$ if $S_i \in \mathcal{S}$, $1 \leq i \leq k$, and $S_1 \cup S_2 \cup \cdots \cup S_k = \mathcal{U}$.

Similarly, $S_1, S_2, \ldots, S_k$ is a k-*partition* of $(\mathcal{U}, \mathcal{S})$ if $S_i \in \mathcal{S}$, $1 \leq i \leq k$, and $S_1 \cup S_2 \cup \cdots \cup S_k = \mathcal{U}$, and $S_i \cap S_j = \emptyset$ for all $i \neq j$. Thus in a k-cover sets S_i and S_j may overlap while in a k-partition all sets are pairwise disjoint. Of course, each k-partition is also a k-cover.

Graph coloring, defined in Sect. 3.1.2, is a nice example of k-partitions and k-covers. Let us recall that a feasible coloring of an undirected graph $G = (V, E)$ assigns a color to each vertex of G such that each pair of adjacent vertices has different colors. The smallest number of colors in a feasible coloring of G is called the chromatic number of G.

Let us consider graph coloring via k-partitions and k-covers. The set of vertices of a graph G obtaining the same color under a feasible coloring, called a color class, is an independent set of G. Hence a coloring using k colors, called a k-*coloring*, corresponds to a partition of V into k independent sets. Thus a k-coloring is a k-partition of $(V, \mathcal{I})$ where V is the vertex set of the graph G and $\mathcal{I}$ is the set of all independent sets of G. On the other hand, given a k-cover of $(V, \mathcal{I})$ it is easy to derive a k-coloring of G. If $I_1, I_2, \ldots, I_k$ is a k-cover of $(V, \mathcal{I})$, then a k-partition of $(V, \mathcal{I})$ and hence, a k-coloring of G can be obtained by simply removing from I_2 the vertices of I_1, from I_3 the vertices of $I_1 \cup I_2$, etc. Therefore, to decide whether G has a k-coloring, it is sufficient to decide whether $(V, \mathcal{I})$ has a k-cover (or k-partition). Actually, by making use of inclusion-exclusion it is possible not only to decide whether $(V, \mathcal{I})$ has a k-cover (k-partition), but also to count the number of k-covers (k-partitions) of $(V, \mathcal{I})$.

4.3.1 Coverings and Graph Coloring

The main goal of this subsection is to present and analyse an inclusion-exclusion algorithm to count the number of k-covers of an input $(\mathcal{U}, \mathcal{S})$. We denote by $c_k = c_k(\mathcal{S})$ the number of ordered k-covers of $(\mathcal{U}, \mathcal{S})$. Hence $S_1, S_2, \ldots, S_k$ and $S_{\pi(1)}, S_{\pi(2)}, \ldots, S_{\pi(k)}$ are considered to be different k-covers for any permutation π different from the identity. Note that the number of unordered k-covers of $(\mathcal{U}, \mathcal{S})$ is $c_k(\mathcal{S})/k!$.

For $W \subseteq \mathcal{U}$, we define

$$\mathcal{S}[W] = \{S \in \mathcal{S} : S \cap W = \emptyset\}.$$

In other words, $\mathcal{S}[W]$ is the subfamily of $\mathcal{S}$ avoiding W. We denote by $s[W]$ the cardinality of $\mathcal{S}[W]$.

First we present an inclusion-exclusion formula for c_k the number of ordered k-covers of $(\mathcal{U}, \mathcal{S})$. Note that the number of ordered k-covers of $(\mathcal{U}, \mathcal{S})$ is the number of ways to choose $S_1, S_2, \ldots, S_k \in \mathcal{S}$ with replacement (also called a k-permutation of $\mathcal{S}$ with repetition) such that $S_1 \cup S_2 \cup \cdots \cup S_k = \mathcal{U}$.

Lemma 4.7. *The number of k-covers of a set system $(\mathcal{U}, \mathcal{S})$ is*

$$c_k = \sum_{W \subseteq \mathcal{U}} (-1)^{|W|} s[W]^k. \tag{4.5}$$

Proof. We show how to establish the formula (4.5) using Theorem 4.2. Using the terminology of Theorem 4.2, we first define the objects. Each object is an ordered set $O = (S_1, S_2, \ldots, S_k)$, $S_i \in \mathcal{S}$. Let us remark that we do not exclude that $S_i = S_j$ for some i and j. Thus the number of objects is equal to $|\mathcal{S}|^k$, which is the number of ways of choosing $S_1, S_2, \ldots, S_k \in \mathcal{S}$ with replacements.

Now we have to define the properties that our objects may or may not have. We say that for $u \in \mathcal{U}$ an object $O = (S_1, S_2, \ldots, S_k)$ has property $Q(u)$ if $u \in \cup_{i=1}^n S_i$. Every object having all properties $Q(u)$ for all $u \in \mathcal{U}$ corresponds to a k-cover because $\cup_{j=1}^n S_j = \mathcal{U}$. Thus X, which is the number of objects having all properties $Q(u)$ for all $u \in \mathcal{U}$, is equal to c_k. Finally, to use Theorem 4.2, we observe that $N(W)$, the number of objects that do not have any of the properties $Q(w)$ for $w \in W$, is equal to $s[W]^k$. (This is just the number of ways one can choose sets $S_1, S_2, \ldots, S_k \in \mathcal{S}[W]$ with replacements.) By Theorem 4.2,

$$X = \sum_{W \subseteq \{1,2,\ldots,n\}} (-1)^{|W|} N(W)$$

and the lemma follows.

$\square$

Theorem 4.8. *There is an algorithm computing $c_k(\mathcal{S})$, the number of (ordered) k-covers of a set system $(\mathcal{U}, \mathcal{S})$ with n elements, in time $\mathcal{O}^*(2^n)$ and exponential space.*

Proof. The algorithm computing c_k is based on the inclusion-exclusion formula of Lemma 4.7 and works as follows.

1. First it builds a table with 2^n entries containing $s[W]$, the number of sets of $\mathcal{S}$ avoiding W, for all $W \subseteq \mathcal{U}$;
2. Then it evaluates (4.5).

The second part of the algorithm is easy. In fact, it is essentially the same for any inclusion-exclusion algorithm based on such an inclusion-exclusion formula. This evaluation by computing the partial sums iteratively can easily be done in time $\mathcal{O}^*(2^n)$. Since the running time of the second part of the algorithm is $\mathcal{O}^*(2^n)$, it is sufficient to show how to construct the required table of $s[W]$ for all $W \subseteq \mathcal{U}$ in time $\mathcal{O}^*(2^n)$. We use dynamic programming to construct the table.

The value $s[W]$ can be expressed as

$$s[W] = \sum_{S \subseteq \mathcal{U} \setminus W} f(S),$$

where f is the *characteristic function*

$$f(S) = \begin{cases} 1, & \text{if } S \in \mathcal{S}, \\ 0, & \text{otherwise.} \end{cases}$$

Let $u_1, u_2, \ldots, u_n$ be any ordering of the elements of $\mathcal{U}$. For $i \in \{0, \ldots, n\}$ and $W \subseteq \mathcal{U}$, we define $g_i(W)$ as the number of subsets of $\mathcal{S}$ avoiding W, and containing all elements from $\{u_{i+1}, \ldots, u_n\} \setminus W$. In other words,

$$g_i(W) = \sum_{\mathcal{U} \setminus (W \cup \{u_1, \ldots, u_i\}) \subseteq S \subseteq \mathcal{U} \setminus W} f(S).$$

By the definition of g_i, we have that

$$g_0(W) = f(\mathcal{U} \setminus W)$$

and

$$g_n(W) = \sum_{S \subseteq \mathcal{U} \setminus W} f(S) = s[W].$$

To compute $g_n(W)$, we observe that for $i \geq 1$,

$$g_i(W) = \begin{cases} g_{i-1}(W) & \text{if } u_i \in W, \\ g_{i-1}(W \cup \{i\}) + g_{i-1}(W) & \text{otherwise.} \end{cases}$$

Indeed, if $u_i \in W$ then

$$\mathcal{U} \setminus (W \cup \{u_1, \ldots, u_i\}) = \mathcal{U} \setminus (W \cup \{u_1, \ldots, u_{i-1}\}).$$

If $u_i \notin W$, then $g_{i-1}(W)$ counts all sets from $g_i(W)$ containing u_i, and $g_{i-1}(W \cup \{u_i\})$ all sets that do not contain u_i.

Thus computation of the values $g_n(W) = s[W]$, for all $W \subseteq \mathcal{U}$, can be done in time $\mathcal{O}^*(2^n)$. □

In Chap. 7, we will see how the ideas from Theorem 4.8 can be used to obtain algorithm for fast zeta and Möbius transforms. The algorithm described in Theorem 4.8 requires exponential space. It is an interesting open problem whether the running time $\mathcal{O}^*(2^n)$ can be achieved by a polynomial space algorithm.

The inclusion-exclusion algorithm to count the number of k-covers can be used to establish an algorithm solving the coloring problem. Notice that the chromatic number of a graph $G = (V, E)$ is the smallest k for which $c_k(V, \mathcal{I}) > 0$, where $\mathcal{I}$ is the collection of all independent sets of the input graph $G = (V, E)$.

Theorem 4.9. *There is an algorithm computing the chromatic number of a graph in time $\mathcal{O}^*(2^n)$ and exponential space.*

Theorem 4.9 is a big improvement on the previously best known algorithms. Lawler's dynamic programming coloring algorithm (see Chap. 3) has running time $\mathcal{O}(2.4423^n)$ and had only been marginally improved before 2006 by refined dynamic programming algorithms.

4.3.2 Partitions

In this subsection we present an inclusion-exclusion algorithm to count k-partitions which is strongly related to the algorithm counting k-coverings in the previous subsection.

Let us recall the problem to be solved. Let $\mathcal{U}$ be a set of n elements and let $\mathcal{S}$ be a collection of subsets of $\mathcal{U}$. We assume that $\mathcal{S}$ can be enumerated in time $\mathcal{O}^*(2^n)$. Now a collection of nonempty subsets $S_1, S_2, \ldots, S_k$ is said to be a k-partition of $(\mathcal{U}, \mathcal{S})$ if $S_i \in \mathcal{S}$ for all $i \in \{1, 2, \ldots, k\}$, $S_1 \cup S_2 \cup \cdots \cup S_k = U$ and $S_i \cap S_j = \emptyset$ for all $i \neq j$. We denote by $p_k = p_k(\mathcal{S})$ the number of unordered k-partitions of $(\mathcal{U}, \mathcal{S})$ and thus $S_1, S_2, \ldots, S_k$ and $S_{\pi(1)}, S_{\pi(2)}, \ldots, S_{\pi(n)}$ are considered to be the same partition for any permutation π. Note that the number of ordered k-partitons of $(\mathcal{U}, \mathcal{S})$ is $k! \cdot p_k(\mathcal{S})$. Our goal is to present an inclusion-exclusion algorithm to compute $p_k(\mathcal{S})$.

We start with an inclusion-exclusion formula for $p_k = p_k(\mathcal{S})$, the number of unordered k-partitions $S_1, S_2, \ldots, S_k \in \mathcal{S}$ of $(\mathcal{U}, \mathcal{S})$.

Lemma 4.10. *The number of k-partitons of a set system $(\mathcal{U}, \mathcal{S})$ is*

$$p_k = \sum_{W \subseteq \mathcal{U}} (-1)^{|W|} a_k[W], \qquad (4.6)$$

where $a_k[W]$ denotes the number of combinations with repetition of k sets $S_1, S_2, \ldots,$ $S_k \in \mathcal{S}[W]$ satisfying $|S_1| + |S_2| + \cdots + |S_k| = n$.

Proof. The proof is based on Theorem 4.2. We choose as combinatorial objects all combinations with repetition $S_1, S_2, \ldots, S_k$ of k sets of $\mathcal{S}$ satisfying $|S_1| + |S_2| + \cdots + |S_k| = n$. Such an object $S_1, S_2, \ldots, S_k$ has property $Q(u)$ for some $u \in \mathcal{U}$ if $u \in \cup_{j=1}^k S_j$.

Thus an object $S_1, S_2, \ldots, S_k$ has property $Q(u)$ for all $u \in \mathcal{U}$ iff $\cup_{j=1}^k S_j = \mathcal{U}$, and therefore, because $|S_1| + |S_2| + \cdots + |S_k| = n$, these are precisely the k-partitions. Finally $N(W)$ is the number of those objects having none of the properties $Q(w)$ with $w \in W$. Hence these are the objects $S_1, S_2, \ldots, S_k$ satisfying $\cup_{j=1}^k S_j \cap W = \emptyset$, which is the case iff $S_1, S_2, \ldots, S_k$ is a combination of k sets of $\mathcal{S}[W]$. Now when substituting all this into the formula (4.2), we obtain (4.6). □

Similarly to the previous section, an algorithm to compute $p_k = p_k(\mathcal{S})$ based on the inclusion-exclusion formula of Lemma 4.10 works as follows:

1. First it builds a table with 2^n entries containing $a_k[W]$ for all $W \subseteq \mathcal{U}$.
2. Then it evaluates (4.6).

The evaluation of the formula (4.6) is done by iteratively computing the partial sums which can easily be done in time $\mathcal{O}^*(2^n)$ and needs only polynomial space.

The crucial part of the inclusion-exclusion algorithm to count the k-partitions of $(\mathcal{U}, \mathcal{S})$ is the computation of $a_k[W]$ for all $W \subseteq \mathcal{U}$. This is done in two steps, both of them using a dynamic programming algorithm.

For $i = 0, 1, 2, \ldots n$, we denote by $\mathcal{S}^{(i)}[W]$ the collection of all sets $S \in \mathcal{S}[W]$ of cardinality i, and we denote by $s^{(i)}[W]$ the number of sets in $\mathcal{S}^{(i)}[W]$. In the first step the algorithm computes $s^{(i)}[W]$ for all $W \subseteq \mathcal{U}$ and all $i \in \{1, 2, \ldots, n\}$. Note that these are $n2^n$ values in total. This is a ranked version of the $\mathcal{O}^*(2^n)$ time algorithm of the previous section which computes $s[W]$ for all $W \subseteq \mathcal{U}$. Simply applying this algorithm to the set system $(\mathcal{U}, \mathcal{S}^{(i)})$ for any fixed $i \in \{1, 2, \ldots, n\}$ computes $s^{(i)}[W]$ for all $W \subseteq U$. Thus, since for every $i \in \{1, 2, \ldots, n\}$, $\mathcal{S}^{(i)}$ can be enumerated in time $\mathcal{O}^*(2^n)$, the values $s^{(i)}[W]$ for all $W \subseteq \mathcal{U}$ and all $i \in \{1, 2, \ldots, n\}$ can be computed in time $\mathcal{O}^*(2^n)$.

In the second step a dynamic programming algorithm computes $a_k[W]$ for all $W \subseteq \mathcal{U}$ using the tables with entries $s^{(i)}[W]$ for all $i \in \{0, 1, \ldots, n\}$ and all $W \subseteq \mathcal{U}$. Let $A(\ell, m, W)$ denote the number of permutations with repetition of ℓ sets $S_1, \ldots, S_\ell \in \mathcal{S}[W]$ satisfying $|S_1| + \cdots + |S_\ell| = m$. Thus $A(k, n, W)$ is the number of ordered k-partitions avoiding W, and therefore $a_k[W] = A(k, n, W)/k!$ since $a_k[W]$ is the number of unordered k-partitions avoiding W.

To compute $A(\ell, m, W)$ for all $\ell = 1, 2, \ldots, k$, for all $m = 1, 2, \ldots, n$ and all $W \subseteq \mathcal{U}$ we use a dynamic programming algorithm based on the following recurrence

$$A(1, m, W) = s^{(m)}[W] \tag{4.7}$$

$$A(\ell, m, W) = \sum_{i=1}^{m-1} s^{(m-i)}[W] \cdot A(\ell - 1, i, W). \tag{4.8}$$

Clearly the dynamic programming algorithm of the second step runs in time $\mathcal{O}^*(2^n)$ and needs exponential space. It computes $a_k[W]$ for all $W \subseteq \mathcal{U}$. Hence p_k can be obtained by evaluating (4.6).

Theorem 4.11. *There is an ialgorithm computing $p_k(\mathcal{S})$, the number of (unordered) k-partitions of a set system $(\mathcal{U}, \mathcal{S})$, in time $\mathcal{O}^*(2^n)$ and exponential space.*

The approach can also be applied to various other NP-hard graph problems as e.g. partition into Hamiltonian subgraphs, partition into forests and partition into matchings.

Exercise 4.12. In the DOMATIC NUMBER problem we are given an undirected graph $G = (V, E)$. The task is to compute the domatic number of G which is the largest integer k such that there is a partition of V into pairwise disjoint sets $V_1, V_2, \ldots V_k$ such that $V_1 \cup V_2 \cup \cdots \cup V_k = V$ and each V_i is a dominating set of G. Construct an $\mathcal{O}^*(2^n)$ time algorithm to compute the domatic number of the given graph.

In Chap. 7 we show how to count k-partitions by a $\mathcal{O}^*(2^n)$ algorithm based on subset convolution. This algorithm as well as the one in this subsection needs exponential space. Polynomial space algorithms to compute c_k and p_k of a given set system $(\mathcal{U}, \mathcal{S})$ will be studied in the next subsection.

4.3.3 Polynomial Space Algorithms

The inclusion-exclusion algorithms to compute c_k and p_k in running time $\mathcal{O}^*(2^n)$, presented in the preceeding subsections, need exponential space. For both algorithms it is necessary that $\mathcal{S}$ can be enumerated in time $\mathcal{O}^*(2^n)$ and exponential space.

To establish polynomial space algorithms to compute c_k and p_k it is necessary that $\mathcal{S}$ can be enumerated in polynomial space. Then the inclusion-exclusion formulas of Sects. 4.3.1 and 4.3.2 imply polynomial space algorithms of running time $\mathcal{O}^*(c^n)$ with $c \leq 4$ for computing c_k and p_k. The following theorem nicely summarizes the running times of polynomial space algorithms to be established.

Theorem 4.13. *The number of k-covers c_k and k-partitions p_k of $(\mathcal{U}, \mathcal{S})$ can be computed in polynomial space and*

(i) $2^n |\mathcal{S}| n^{\mathcal{O}(1)}$ *time, assuming that $\mathcal{S}$ can be enumerated in polynomial space with polynomial time delay*
(ii) $3^n n^{\mathcal{O}(1)}$ *time, assuming membership in $\mathcal{S}$, i.e. "$S \in \mathcal{S}$?", can be decided in polynomial time, and*
(iii) $\sum_{j=0}^{n} \binom{n}{j} T_{\mathcal{S}}(j)$ *time, assuming that there is a $T_{\mathcal{S}}(j)$ time and polynomial space algorithm to count for any j-element subset $W \subseteq \mathcal{U}$ the number of sets $S \in \mathcal{S}$ satisfying $S \cap W = \emptyset$.*

Proof. First we show how to compute c_k. The exponential space algorithm given in Sect. 4.3.1 is based on Lemma 4.7 and the inclusion-exclusion formula (4.5)

$$c_k = \sum_{W \subseteq \mathcal{U}} (-1)^{|W|} s[W]^k,$$

where $s[W]$ is the number of $S \in \mathcal{S}$ not intersecting W.

The polynomial space algorithm to compute c_k iterates over all $W \subseteq \mathcal{U}$, adding the value of $(-1)^{|W|} s[W]^k$ to a running total sum. The difficulty is to compute $s[W]$ which is done separately for all $W \subseteq \mathcal{U}$. We distinguish three cases.

(i)   For each $W \subseteq \mathcal{U}$, the polynomial delay and polynomial space enumeration algorithm is used to enumerate $\mathcal{S}$ in time $\mathcal{O}^*(|\mathcal{S}|)$, and to verify for each $S \in \mathcal{S}$ whether $S \cap W = \emptyset$, and thus for any W the value of $s[W]$ can be computed in time $\mathcal{O}^*(|\mathcal{S}|)$. Therefore the overall running time of the algorithm computing c_k is $2^n |\mathcal{S}| n^{\mathcal{O}(1)}$.
(ii) For each $W \subseteq \mathcal{U}$, the algorithm tests all $2^{n-|W|}$ subsets S of $\overline{W}$ for membership in $\mathcal{S}$, each one in polynomial time. This amounts to a total running time of $\sum_{i=1}^{n} \binom{n}{n-i} 2^{n-i} n^{\mathcal{O}(1)} = \mathcal{O}^*(3^n)$.
(iii) By assumption, there is a polynomial space algorithm to compute $s[W]$ in time $T_{\mathcal{S}}(j)$ for every $W \subseteq \mathcal{U}$ with $|W| = j$. Using this algorithm for any W, the total running time of the inclusion-exclusion algorithm to compute c_k is $\sum_{j=0}^{n} \binom{n}{j} T_{\mathcal{S}}(j)$.

Now we show how to compute p_k. The algorithm given in Sect. 4.3.2 is based on the inclusion-exclusion formula (4.6)

$$p_k = \sum_{W \subseteq \mathcal{U}} (-1)^{|W|} a_k(W),$$

where $a_k(W)$ denotes the number of ways to choose k sets $S_1, S_2, \ldots, S_k \in \mathcal{S}[W]$, possibly overlapping, such that $|S_1| + |S_2| + \cdots + |S_k| = n$. The exponential space algorithm comprises two dynamic programming algorithms. First for all $W \subseteq \mathcal{U}$ and all $i = 0, 1, \ldots n$, the number of sets $S \in \mathcal{S}[W]$ of cardinality i, denoted by $s^{(i)}[W]$, is computed. Based on the values of $s^{(i)}[W]$ for all $W \subseteq \mathcal{U}$ and all $i \in \{0, 1, \ldots, n\}$, the second dynamic programming algorithm computes $a_k(W)$ for every $W \subseteq \mathcal{U}$ based on the recurrences (4.7) and (4.8).

The polynomial space algorithm, like the exponential space algorithm, iterates over all $W \subseteq \mathcal{U}$, adding the value of $(-1)^{|W|} a_k(W)$ to a running total sum. The difficulty is to compute $a_k(W)$ for all $W \subseteq \mathcal{U}$ in polynomial space. Fortunately the second dynamic programming algorithm based on recurrences (4.7) and (4.8) can be done separately for each $W \subseteq \mathcal{U}$, and is thus a polynomial space algorithm with running time $\mathcal{O}^*(2^n)$. Hence it remains to show how to compute $s^{(i)}[W]$ for all $W \subseteq \mathcal{U}$ and all $i \in \{0, 1, \ldots, n\}$ in polynomial space. We distinguish three cases. Note that in all cases the computation is done separately for different $W \subseteq \mathcal{U}$.

(i)   For every $W \subseteq \mathcal{U}$ and all $i \in \{0, 1, \ldots, n\}$, the polynomial delay and polynomial space enumeration algorithm is used to enumerate $\mathcal{S}$ in time $\mathcal{O}^*(|\mathcal{S}|)$, and to verify for each $S \in \mathcal{S}$ with $|S| = i$ whether $S \cap W = \emptyset$, and thus for any W, the value of $s^{(i)}[W]$ can be computed in time $\mathcal{O}^*(|\mathcal{S}|)$. Therefore the overall running time of the algorithm computing c_k is $2^n |\mathcal{S}| n^{\mathcal{O}(1)}$.

(ii)   For every $W \subseteq \mathcal{U}$ and all $i \in \{0, 1, \ldots, n\}$, the algorithm tests altogether all $2^{n-|W|}$ subsets S of $\overline{W}$ for membership in $\mathcal{S}$, each one in polynomial time. This amounts to a total running time of $\sum_{i=1}^{n} \binom{n}{n-i} 2^{n-i} n^{\mathcal{O}(1)} = \mathcal{O}^*(3^n)$.

(iii)   By assumption, there is a $T_\mathcal{S}(j)$ time and polynomial space algorithm to count for any j-element subset $W \subseteq U$ the number of sets $S \in \mathcal{S}^{(i)}$ satisfying $S \cap W = \emptyset$, which is precisely $s^{(i)}[W]$. Using this algorithm, for any $W \subseteq \mathcal{U}$ and all $i \in \{0, 1, \ldots, n\}$, the total running time of the inclusion-exclusion algorithm to compute p_k is $\sum_{j=0}^{n} \binom{n}{j} T_\mathcal{S}(j)$.

$\square$

Theorem 4.13 provides a framework for designing polynomial space algorithms for covering and partition problems. Let us reconsider the graph coloring problem for which an exponential space algorithm is given in Chap. 3.

Corollary 4.14. *The chromatic number of a graph can be computed by a polynomial space algorithm in time* $\mathcal{O}(2.2461^n)$.

Proof. The polynomial space algorithm to compute the chromatic number is based on a polynomial space algorithm to compute $s[W]$ for all $W \subseteq U$, which is an algorithm to count all independent sets S in the graph $G \setminus W$. This is a polynomial space $\mathcal{O}(1.2461^n)$ time algorithm to count the number of independent sets in an n-vertex graph [99]. By Theorem 4.13 (iii), this implies a polynomial space algorithm to compute $c_k(\mathcal{I})$ of running time

$$\sum_{j=0}^{n} \binom{n}{n-j} O(1.2461^j) = O(2.2461^n).$$

Thus the chromatic number can be computed in time $\mathcal{O}(2.2461^n)$.

$\square$

Exercise 4.15. Find a polynomial space algorithm to compute the domatic number of a given graph.

4.4 Counting Subgraph Isomorphisms

The HAMILTONIAN PATH problem studied in previous sections is a special case of a more general problem, namely the SUBGRAPH ISOMORPHISM problem.

In the SUBGRAPH ISOMORPHISM problem we are given two (directed or undirected) graphs $F = (V_F, E_F)$ and $G = (V_G, E_G)$. The question is whether G contains a copy of F, or in other words, whether G has a subgraph isomorphic to F. More formally, we say that a graph G has a subgraph $H = (V_H, E_H)$ isomorphic to $F = (V_F, E_F)$, if there is a bijection $f: V_F \to V_H$ such that for every pair of vertices $u, v \in V_F$, the vertices $f(u)$ and $f(v)$ are adjacent in H if and only if $\{u, v\}$ is an edge of F.

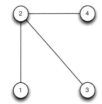 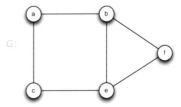

Fig. 4.2 Graph G has two subgraphs isomorphic to F: the graph formed by the edges $\{a,b\}$, $\{b,e\}$, $\{b,f\}$, and the graph formed by edges $\{c,e\}$, $\{b,e\}$, $\{e,f\}$.

To give an example, if F is a (chordless) cycle on $n = |V_G|$ vertices, then the question whether G has a subgraph isomorphic to F is equivalent to the question if G contains a cycle passing through all its vertices, i.e. whether G has a Hamiltonian cycle. Another example is given in Fig. 4.2.

While COLORING is not a subgraph isomorphism problem, it can be reduced to SUBGRAPH ISOMORPHISM. Let $\overline{G}$ be the complement of G, i.e. the graph on the same vertex set as G, and for every pair $u \neq v$ of vertices, $\{u, v\}$ is an edge in $\overline{G}$ if and only if $\{u,v\}$ is not an edge of G. Then G can be colored in k colors if and only if the vertices of $\overline{G}$ can be covered by k disjoint cliques. In other words, $\chi(G) \leq k$ if and only if $\overline{G}$ contains as subgraph a graph F which is a disjoint union of cliques $K_{t_1}, K_{t_2}, \cdots, K_{t_k}$, such that $t_1 + t_2 + \cdots + t_k = n$. See Fig. 4.3.

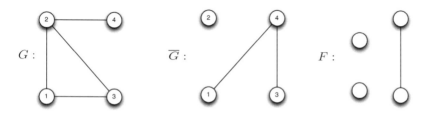

Fig. 4.3 The complement $\overline{G}$ of G contains a subgraph isomorphic to F, which is a disjoint union of cliques K_1, K_1, and K_2. Moreover, the number of vertices in F is equal to $1 + 1 + 2$, the number of vertices in G. Thus G can be colored with 3 colors.

In this section we discuss algorithms for counting the number of copies of a given graph F in a graph G, or in other words, counting the subgraphs isomorphic to F in G. This counting problem becomes hard even for very simple graphs F. For example, when the number of vertices of G is $2n$, and the graph F is a disjoint union of n edges, then the number of subgraphs of G isomorphic to F is exactly the number of perfect matchings in G. Counting perfect matchings in a graph is a well known #P-complete problem.

To count perfect matching, Hamiltonian paths, and colorings in this chapter, we counted objects that were easier to deal with, and then used inclusion-exclusion. The same strategy can be applied to count isomorphic subgraphs, and the role of "simple" objects is given to graphs with "easily" computable homomorphisms. From this perspective, graph homomorphism is a generic approach to several algorithmic results including Theorems 4.4, 4.5, and 4.9. In what follows, we explain how graph homomorphisms can be used to prove Theorems 4.5 and 4.9. In Sect. 5.3, we use the ideas from this section about graph homomorphisms to count more complicated subgraphs in a graph than cycles and cliques.

Let $F = (V_F, E_F)$ and $G = (V_G, E_G)$ be two (directed or undirected) graphs. A *homomorphism* from F to G is a mapping f from V_F to V_G, that is $f : V_F \rightarrow V_G$, such that if $\{u,v\} \in E_F$ then $\{f(u), f(v)\} \in E_G$. Furthermore, if the mapping f is injective then f is called an *injective homomorphism*. Let us note that there is an injective homomorphism from F to G if an only if G contains a subgraph isomorphic to F.

For example, for the graphs in Fig. 4.2, the mapping f with $f(2) = a$, $f(1) = f(3) = f(4) = b$ is a homomorphism from F to G. Furthermore the mapping g with $g(1) = a$, $g(2) = b$, $g(3) = e$, and $g(4) = f$ is an injective homomorphism from F to G. Let us note that in the case of an injective homomorphism, the image of F in G is isomorphic to F. For another example, if f is a homomorphism from F

to G, and F is a clique on p vertices, then $f(F)$ is also a clique on p vertices in G. If $F = (v_1, v_2, \ldots, v_p)$ is a path of length $p-1$, then $(f(v_1), f(v_2), \ldots, f(v_p))$ is a walk of length $p-1$ in G. If f is an injective homomorphism from the path $F = (v_1, v_2, \ldots, v_p)$ to G, then the image of F in G is also a path of length $p-1$.

We use $\hom(F, G)$, $\text{inj}(F, G)$ and $\text{sub}(F, G)$ to denote the number of homomorphisms from F to G, the number of injective homomorphisms from F to G and the number of distinct copies of F in G, respectively. Let us remark that while G has a subgraph isomorphic to F if and only if there is an injective homomorphism from F to G, the numbers $\text{inj}(F, G)$ and $\text{sub}(F, G)$ can be quite different. For example, take the graph F depicted in Fig. 4.2 and the graph $G = F$. Of course, every graph is isomorphic to itself, and G contains exactly one copy of F. However, there are several injective homomorphisms from F to F, also called *automorphisms* of F. For the graph F in Fig. 4.2 there are $3! = 6$ injective homomorphisms. Each of these homomorphisms has as a fixed point vertex 2 but the leaves of the tree are not necessarily fixed.

Let $\text{aut}(F)$ be the number of automorphisms of F, i.e. the number of injective homomorphisms from F to F. The values $\text{inj}(F, G)$ and $\text{sub}(F, G)$ are related to each other by the following proposition which follows directly from the definitions of $\text{inj}(F, G)$, $\text{sub}(F, G)$, and $\text{aut}(F)$.

Proposition 4.16. *Let F and G be two graphs. Then*

$$\text{sub}(F, G) = \frac{\text{inj}(F, G)}{\text{aut}(F)}.$$

The following is a general result relating the number of injective homomorphisms and the number of homomorphisms from F to G using an inclusion-exclusion formula.

Theorem 4.17. *Let $F = (V_F, E_F)$ and $G = (V_G, E_G)$ be two graphs with $|V_G| = |V_F|$. Then*

$$\text{inj}(F, G) = \sum_{W \subseteq V_G} (-1)^{|W|} \hom(F, G \setminus W).$$

Proof. Let us see how to obtain the proof of the theorem from Theorem 4.2. Using the notation of Theorem 4.2, the set of objects is the set of homomorphisms from F to G. For a homomorphism f and a vertex $v \in V_G$ we define the property $Q(v)$ as the property that v does not belong to the image $f(V_F)$. In other words, f has property $Q(v)$ if $v \notin \bigcup_{x \in V_F} f(x)$. Then for a subset $W \subseteq V_G$, $N(W)$ is the number of homomorphism f such that the image $f(V_F)$ does not contain a vertex from W. Thus $N(W) = \hom(F, G \setminus W)$. The number X is the number of homomorphisms possessing none of the properties $Q(v)$, $v \in V_G$. But this is exactly the number of injective homomorphisms and thus $X = \text{inj}(F, G)$. Now the theorem follows from (4.2). $\qquad\qquad\square$

How to use Theorem 4.17 algorithmically? Assume that for some graph F we can compute the number of graph homomorphisms from F to all the graphs $G[W]$ in time $t(n)$, where $|V_F| \leq |V_G|$ and $W \subseteq V_G$. Then, as a consequence of Theorem 4.17 , we can compute the value of $\text{inj}(F, G)$ in time $\mathcal{O}(2^n \cdot t(n))$ when $|V_F| = |V_G| = n$. A natural question arising here is to extend this to the case when the size of V_F, say n_F, is less than $n = |V_G|$. The easiest solution will be to enumerate all subsets V' of size n_F of V_G and then to compute $\text{inj}(F, G[V'])$. But this will take time $\mathcal{O}(\binom{n}{n_F} 2^{n_F} t(n))$, which in the worst case, is $\mathcal{O}(3^n \cdot t(n))$. In the next theorem we show how to extend Theorem 4.17 to the case $n_F < n$.

Theorem 4.18. *Let $F = (V_F, E_F)$ and $G = (V_G, E_G)$ be two graphs with $|V_F| = n_F \leq |V_G| = n$. Then*

$$\text{inj}(F, G) = \sum_{Y \subseteq V_G, |Y| \leq n_F} (-1)^{n_F - |Y|} \binom{n - |Y|}{n_F - |Y|} \text{hom}(F, G[Y]).$$

Proof. By Theorem 4.17,

$$\text{inj}(F, G) = \sum_{W \subseteq V_G} (-1)^{|V| - |W|} \text{hom}(F, G[W]). \tag{4.9}$$

Now

$$\begin{aligned}
\text{inj}(F, G) &= \sum_{W \subseteq V(G), |W| = n_F} \text{inj}(F, G[W]) \\
&\stackrel{\text{by (4.9)}}{=} \sum_{W \subseteq V_G, |W| = n_F} \left(\sum_{Y \subseteq W} (-1)^{|W| - |Y|} \text{hom}(F, G[Y]) \right) \\
&= \sum_{W \subseteq V_G, |W| = n_F} \left(\sum_{Y \subseteq W} (-1)^{n_F - |Y|} \text{hom}(F, G[Y]) \right) \\
&= \sum_{Y \subseteq V_G, |Y| \leq n_F} (-1)^{n_F - |Y|} \binom{n - |Y|}{n_F - |Y|} \text{hom}(F, G[Y]).
\end{aligned}$$

The last equality follows from the fact that for any subset Y with $|Y| \leq n_F$, the value of $\text{hom}(F, G[Y])$ is counted precisely for all those subsets W for which $Y \subseteq W$ and $|W| = n_F$. On the other hand, for every fixed Y, $\text{hom}(F, G[Y])$ is counted once in the above sum for every superset W of Y of size n_F. The number of such sets W is precisely $\binom{n - |Y|}{n_F - |Y|}$. Furthermore, for all such sets, we have the same sign corresponding to Y, that is, $(-1)^{n_F - |Y|}$. This completes the proof. $\square$

In what follows, we give two examples of the use of graph homomorphisms: Counting Hamiltonian cycles and counting colorings. We come back to graph homomorphisms with applications to graphs of bounded treewidth in Chap. 5.

Let $\#\text{HAM}(G)$ denote the number of Hamiltonian cycles in a graph $G = (V, E)$, $|V| = n$, and $F = C_n$ be a cycle of length n. To compute $\text{sub}(F, G) = \#\text{HAM}(G)$, we want to use Theorem 4.17 with Proposition 4.16. To use them we need to compute $\text{aut}(C_n)$ and $\text{hom}(C_n, G)$.

The cycle C_n can be represented as the 1-skeleton of a regular n-gon in the plane. All its automorphisms corresponds to rotations and reflections of a regular n-gon. We leave the proof of the following fact, that implies $\text{aut}(C_n) = 2n$, as an exercise.

Exercise 4.19. There are n rotations and n reflections of a regular n-gon.

There are several ways of computing $\text{hom}(C_n, G)$. One way is to observe that for every homomorphism the image of C_n in G is a closed walk of length n in G, i.e. the walk $\{v_1, v_2, \ldots, v_n\}$, where $v_1 = v_n$. The number of closed walks of a given length can be computed in polynomial time by dynamic programming exactly as was done in the proof of Theorem 4.5 for (not closed) walks. Another way is to prove that $\text{hom}(C_n, G) = \sum_{i=1}^{n} \lambda_i^n$, where $\lambda_1, \ldots, \lambda_n$ are the eigenvalues of the adjacency matrix of G. The third way of counting $\text{hom}(C_n, G)$ is based on dynamic programming on graphs of bounded treewidth. The treewidth of C_n is two, and $\text{hom}(C_n, G)$ is computable in polynomial time by techniques that will be described in Chap. 5.

Summarizing, $\text{hom}(C_n, G)$ and $\text{aut}(C_n)$ are computable in polynomial time. Thus by Theorem 4.17, the number of Hamiltonian cycles in G can be found in time which is proportional (up to a polynomial factor) to the number of subsets in V. This implies

Theorem 4.20. *The number of Hamiltonian cycles* #HAM(G) *in an n-vertex graph G is computable in time* $\mathcal{O}^*(2^n)$.

We already discussed how the chromatic number of a graph can be computed in time $\mathcal{O}^*(2^n)$. Let us see how (slightly worse) $2^{n+\mathcal{O}(\sqrt{n})}$ running time can be achieved by treating the problem as a special case of the COUNTING SUBGRAPH ISOMOR-PHISMS problem, and thus by counting graph homomorphisms.

We denote by $\chi(G; k)$ the number of k-colorings of a graph G. For example, if P_3 is a path on 3 vertices, then $\chi(P_3; 0) = \chi(P_3; 1) = 0$, $\chi(P_3; 2) = 2$, and $\chi(P_3; 3) = 12$. Another example: for a complete graph K_k, $\chi(K_k; k) = k!$. Then the chromatic number of a graph G is the smallest integer $k > 0$ such that $\chi(G; k) > 0$. It is well known and easy to see that for an integer $k \geq 0$, $\chi(G; k) = \text{hom}(G, K_k)$, where K_k is a complete graph on k vertices. However to construct an exact algorithm, we have to look at homomorphisms from "the other side".

As we already discussed, a k-coloring of a graph $G = (V, E)$ can be viewed as partitioning the vertex set of the given graph into k independent sets, that is, a partition $(V_1, \ldots, V_k)$ of V such that for every $i \in \{1, \ldots, k\}$, the graph $G[V_i]$ does not contain an edge. For our purpose, we reformulate the COLORING problem as a problem of partitioning the vertex set into k cliques in the complement graph. Let $\overline{G} = (V, \overline{E})$ be the complement of G. Then G can be partitioned into k independent sets if and only if $\overline{G}$ can be partitioned into k cliques. We model this as a problem of subgraph isomorphism as follows: we guess the sizes $t_1, t_2, \ldots, t_k$ of these cliques, where $\sum_i t_i = n$. Then $\overline{G}$ can be partitioned into cliques of sizes $t_1, t_2, \ldots, t_k$ respectively if and only if there is a subgraph of $\overline{G}$ isomorphic to

$$F = \cup_{i=1}^{k} K_{t_i}.$$

To compute the value of $\chi(G;k)$ for a given graph G, we count the number of partitions of $\overline{G}$ into k cliques. Indeed, every partition of $\overline{G}$ into k cliques corresponds to $k!$ colorings of G—we select one color for all vertices of the same clique, and the number of different choices is $k!$. Let $\mathcal{P}_k(n)$ be the set of all *unordered partitions* of n into k parts. For every partition $\zeta = (t_1, t_2, \ldots, t_k) \in \mathcal{P}_k(n)$, let $F(\zeta) = \cup_i^k K_{t_i}$. Then

$$\chi(G;k) = \sum_{\zeta \in \mathcal{P}_k(n)} k! \cdot \text{sub}(F(\zeta), \overline{G}). \qquad (4.10)$$

Indeed, every subgraph of $\overline{G}$ isomorphic to $F(\zeta)$ corresponds to a partition of $\overline{G}$ into cliques, which is a partition of G into independent sets. But every partition into k independent sets corresponds to $k!$ different colorings.

In order to obtain the desired running time, we need a classical result from number theory giving an upper bound on the number of unordered partitions of n into k parts. Let $p(n)$ be the partition function, i.e. the number of partitions of n. The asymptotic behavior of $p(n)$ was given by Hardy and Ramanujan in their paper in which they develop the famous "circle method" [110].

Theorem 4.21. $p(n) \sim e^{\pi\sqrt{\frac{2n}{3}}}/4n\sqrt{3}$, *as* $n \to \infty$.

Furthermore one can give an algorithm listing all different unordered partitions of n into k parts in time $2^{\mathcal{O}(\sqrt{n})}$. Actually, such a listing is possible even with polynomial (time) delay. This type of algorithm is out of the scope of this book; we refer to the book of Nijenhuis and Wilf for details [165].

Now our strategy to compute $\chi(G;k)$ is as follows. For every partition $\zeta = (t_1, t_2, \ldots, t_k) \in \mathcal{P}_k(n)$ we want to compute the inner sum in (4.10). To compute (4.10), we have to compute the value of $\text{sub}(F(\zeta), \overline{G})$, and to do this we use Theorem 4.17. To implement Theorem 4.17, we have to compute the values of

- $\text{aut}(F(\zeta))$, and
- $\text{hom}(F(\zeta), \overline{G}[V \setminus W])$, where $W \subseteq V$.

The computation of $\text{aut}(F(\zeta))$ is easy—the number of automorphisms of a complete graph on t vertices is $t!$. If $F(\zeta)$ consists of several connected components, then every automorphism either maps a component (complete graph) into itself, or to a component of the same size. Let $n(x)$ be the number of components of size x in $F(\zeta)$ and let $x_1, x_2, \ldots, x_p$, $p \leq k$, be the sizes of the components in $F(\zeta)$. Let us note that x_i is not necessarily equal to t_i because it is possible in the partition ζ that for some $i \neq j$, $t_i = t_j$. Then

$$\text{aut}(F(\zeta)) = \prod_{x \in \{x_1, x_2, \ldots, x_p\}} n(x)! x!,$$

and this value is computable in polynomial time for each ζ.

To compute $\text{hom}(F(\zeta), \overline{G}[V \setminus W])$ we observe that it is sufficient to count the homomorphisms from every component of $F(\zeta)$. The following result for a graph F with several connected components is easy to obtain.

Proposition 4.22. *Let $F_1, \ldots, F_\ell$ be the connected components of the graph F. Then* $\hom(F,G) = \prod_{i=1}^{\ell} \hom(F_i, G)$.

Since every component of $F(\zeta)$ is a complete graph, by Proposition 4.22 all we need are the values of $\hom(K_t, \overline{G}[V \setminus W])$. For every homomorphism from K_t to $\overline{G}[V \setminus W])$, the image of the complete graph K_t is a clique and

$$\hom(K_t, \overline{G}[V \setminus W]) = T[V \setminus W][t] t!,$$

where $T[V \setminus W][t]$ is the number of cliques of size t in $\overline{G}[V \setminus W]$.

Thus to finish all computations we have to find the number of cliques of size t in a graph. By making use of dynamic programming over vertex subsets $W \subseteq V$, we compute the numbers $T[W][i]$, which is the number of cliques of size i in $\overline{G}[W]$. Our dynamic programming algorithm is based on the observation that for $i > 0$,

$$T[W][i] = T[W \setminus \{v\}][i] + T[N(v) \cap W][i-1]$$

for some vertex v. Indeed, for every clique of size i containing v, the neighborhood of i contains a clique of size $i - 1$. By making use of this observation, it is straightforward now to compute the values $T[W][i]$ for all $W \subseteq V$ and $0 \le i \le n$ in time $\mathcal{O}(2^n n^2)$ using exponential space.

To conclude, we have shown that for every partition ζ, and all subsets $W \subseteq V$, we are able to compute $\hom(F(\zeta), \overline{G}[V \setminus W])$ in time $\mathcal{O}^*(2^n)$. Thus by Theorem 4.17, $\mathrm{sub}(F(\zeta), \overline{G})$ is computable in time $\mathcal{O}^*(2^n)$. By Theorem 4.21, the number of all possible partitions ζ is $2^{\mathcal{O}(\sqrt{n})}$, and thus by computing $\mathrm{sub}(F(\zeta), \overline{G})$ for each partition, we compute $\chi(G,k)$ in time $2^{n+\mathcal{O}(\sqrt{n})}$ by making use of (4.10).

Theorem 4.23. *The number of k-colorings of a graph G, denoted by $\chi(G,k)$, can be computed in time $2^{n+\mathcal{O}(\sqrt{n})}$.*

Notes

The principle of inclusion-exclusion is one of the basic tools in combinatorics. The origin of the principle is not clear. Sometimes it is attributed to Sylvester. Almost every textbook in combinatorics devotes a chapter to this method. We refer to books of Aigner [2], Cameron, [46], van Lint and Wilson [153], Ryser [194], and Stanley [207] for detailed discussions of the method.

The application of the inclusion-exclusion principle to exact algorithms was rediscovered several times. In 1969, Kohn, Gottlieb, and Kohn [136] used it to obtain exact algorithms for the TRAVELLING SALESMAN problem. It was used in 1982, by Karp [127] and in 1993, by Bax [11, 12]. A generalization of inclusion-exclusion by finite difference sets was proposed by Bax and Franklin in [10].

The book of Minc is devoted to the study of permanents [156]. The formula for computing the permanent of a matrix is due to Ryser [194]. Exact algorithms to compute the permanent of a matrix over rings and finite commutative semirings are given in [128, 215]. The inclusion-exclusion algorithm of this chapter for counting Hamiltonian paths can be generalized to solve the TSP problem on n cities with maximum distance W between two cities to achieve running time $\mathcal{O}^*(W2^n)$ and space $\mathcal{O}^*(W)$. The self-reduction algorithm was communicated to us by Mikko Koivisto.

Björklund and Husfeldt [24] and Koivisto [137] presented the first $\mathcal{O}^*(2^n)$ time algorithm to solve COLORING using inclusion-exclusion. The journal version [30] of both conference papers presents a general approach that solves various partition problems. Our presentation follows [24]. Nederlof [162] and Lokshtanov and Nederlof [154] further developed inclusion-exclusion based techniques to obtain a number of polynomial space algorithms. The approach based on a combination of branching and inclusion-exclusion is discussed in [190]. The use of homomorphisms for solving the COUNTING SUBGRAPH ISOMORPHISMS problem was studied by Amini, Fomin and Saurabh in [5].

Chapter 5
Treewidth

The treewidth of a graph is one of the most fundamental notions in graph theory and graph algorithms. In this chapter, we give several applications of treewidth in exact algorithms. We also provide an exact algorithm computing the treewidth of a graph.

5.1 Definition and Dynamic Programming

There are several equivalent definitions of treewidth. Here we follow the definition of Robertson and Seymour [182]. A tree decomposition of a graph G is a way of representing G as a tree-like structure. In this chapter we will see how the tree decompositions and path decompositions of graphs can be used as tools for exact algorithms.

Tree decomposition. A *tree decomposition* of a graph $G = (V, E)$ is a pair $(\{X_i \mid i \in I\}, T = (I, F))$ with $\{X_i \mid i \in I\}$ a collection of subsets of V, called *bags*, and $T = (I, F)$ a tree, such that

(T1)   For every $v \in V$, there exists $i \in I$ with $v \in X_i$.
(T2)   For every $\{v, w\} \in E$, there exists $i \in I$ with $v, w \in X_i$.
(T3)   For every $v \in V$, the set $I_v = \{i \in I \mid v \in X_i\}$ forms a connected subgraph (subtree) of T.

The *width* of tree decomposition $(\{X_i \mid i \in I\}, T = (I, F))$ equals $\max_{i \in I} |X_i| - 1$. The *treewidth* of a graph G, $\mathrm{tw}(G)$, is the minimum width of a tree decomposition of G.

To distinguish the vertices of the decomposition tree T and the vertices of the graph G, we will refer to the vertices of T as to nodes.

Exercise 5.1. Prove that the treewidth of a tree with at least two vertices is 1, and that the treewidth of a clique on $n \geq 1$ vertices is $n - 1$.

The *path decomposition* of a graph is a tree decomposition with tree T being a path. A path decomposition is often denoted by listing the successive sets

F.V. Fomin, D. Kratsch, *Exact Exponential Algorithms*, Texts in Theoretical
Computer Science. An EATCS Series, DOI 10.1007/978-3-642-16533-7_5,
© Springer-Verlag Berlin Heidelberg 2010

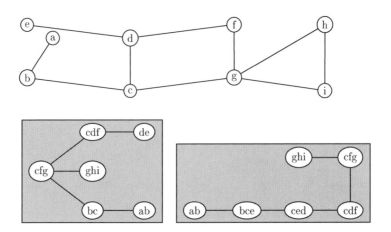

Fig. 5.1 A graph with a tree and a path decomposition

$$(X_1, X_2, \ldots, X_r).$$

Then the *width* of a path decomposition $(X_1, X_2, \ldots, X_r)$ is $\max_{1 \le i \le r} |X_i| - 1$. The *pathwidth* of a graph G, denoted by $\mathrm{pw}(G)$, is the minimum width of a path decomposition of G. Clearly for all graphs G, $\mathrm{tw}(G) \le \mathrm{pw}(G)$.

The property of tree decompositions which is important for performing dynamic programming over tree decompositions is the following.

Exercise 5.2 (Treewidth separator). Let $(\{X_i \mid i \in I\}, T = (I, F))$ be a tree decomposition of a graph G and let i, j, k be nodes of T such that j is on the path from i to k. Then vertex set X_j separates $X_i \setminus X_j$ and $X_k \setminus X_j$, which means that for every pair of vertices $u \in X_i \setminus X_j$ and $v \in X_k \setminus X_j$, every (u,v)-path in G contains a vertex from X_j.

The ideas of dynamic programming on graphs of bounded pathwidth and treewidth are quite similar. However, for pathwidth the description of algorithms is significantly simpler. We start with algorithms on graphs of bounded pathwidth.

It is convenient to work with nice decompositions. A path decomposition

$$(X_1, X_2, \ldots, X_r).$$

of a graph G is *nice* if $|X_1| = |X_r| = 1$, and for every $i \in \{1, 2, \ldots, r-1\}$ there is a vertex v of G such that either $X_{i+1} = X_i \cup \{v\}$, or $X_{i+1} = X_i \setminus \{v\}$. Let us note that because of Property (T3) of tree decompositions, every vertex of G belongs to consecutive sets of bags, and thus the number of bags in a nice path decomposition is at most twice the number of vertices of G. See Fig. 5.2 for an example of a nice path decomposition.

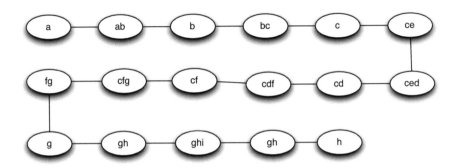

Fig. 5.2 Nice path decomposition of the graph from Fig. 5.1

Exercise 5.3. Let X be a given path decomposition of a graph G of width k. Then G has a nice path decomposition of width k, moreover such a decomposition can be constructed in linear time from X.

We exemplify the way dynamic programming on graphs of bounded pathwidth works on the following NP-complete graph problem.

Maximum Cut. In the MAXIMUM CUT problem (Max-Cut), we are given an undirected graph $G = (V, E)$. For subsets of vertices V_1 and V_2 of G we define $\text{CUT}(V_1, V_2)$ as the number of edges between V_1 and V_2, i.e. the number of edges having one endpoint in V_1 and one endpoint in V_2. The task is to find a set $X \subseteq V$ that maximizes the value of $\text{CUT}(X, V \setminus X)$.

Lemma 5.4. *Let $G = (V, E)$ be a graph on n vertices with a given path decomposition of width at most k. Then the* MAXIMUM CUT *problem on G is solvable in time* $\mathcal{O}(2^k \cdot k \cdot n)$.

Proof. By making use of Exercise 5.3, we transform in linear time the given path decomposition into a nice path decomposition

$$P = (X_1, X_2, \ldots, X_r)$$

of G of width k.

For $i \in \{1, 2, \ldots, r\}$, we put

$$V_i = \bigcup_{j=1}^{i} X_j.$$

Thus $V_r = V$. For every possible partition of X_i into sets (A, B) we want to compute the value $c_i(A, B)$, which is the maximum size of a cut in the graph $G[V_i]$, where the maximum is taken over all cuts (A', B') such that $A \subseteq A'$ and $B \subseteq B'$. In other words, $c_i(A, B)$ is the maximum number of edges in a cut in $G[V_i]$ in a partition of V respecting the partition (A, B). Then the maximum number of edges in a cut in G is

$$\max\{c_r(A,B)\colon (A,B) \text{ is a partition of } X_r\}.$$

We use dynamic programming to compute $c_i(A,B)$ for every partition (A,B) of X_i. The case $i = 1$ is trivial, here $X_1 = v$ for some $v \in V$, and there are only two partitions of X_1, namely $(\{v\},\emptyset)$ and $(\emptyset,\{v\})$. In both cases, $c_1(\{v\},\emptyset) = c_1(\emptyset,\{v\}) = 0$. For $i > 1$, we consider two cases

— $X_i = X_{i-1} \cup \{v\}$, for some $v \in V$. By Exercise 5.2, all neighbors of v in V_i must be in $X_{i-1} \cap X_i$ (for every $u \in V_{i-1} \setminus X_{i-1}$, the set X_{i-1} must separate u and v). Then for every partition (A,B) of X_i,

$$c_i(A,B) = \begin{cases} c_{i-1}(A \setminus \{v\}, B) + CUT(\{v\}, B), & \text{if } v \in A, \\ c_{i-1}(A, B \setminus \{v\}) + CUT(\{v\}, A), & \text{if } v \notin A \end{cases} \qquad (5.1)$$

— $X_i = X_{i-1} \setminus \{v\}$, for some $v \in V$. In this case, for every partition (A,B) of X_i, we have that

$$c_i(A,B) = \max\{c_{i-1}(A \cup \{v\}, B), c_{i-1}(A, B \cup \{v\})\}. \qquad (5.2)$$

The computation of the values $c_i(A,B)$ in (5.1) and (5.2) can be done in time $\mathcal{O}(2^{|X_i|}|X_i|)$ and by making use of space $\mathcal{O}(2^{|X_i|})$. Thus the total running time is

$$\mathcal{O}(\sum_{i=1}^{r} 2^{|X_i|}|X_i|) = \mathcal{O}(2^k \cdot k \cdot n).$$

$\square$

As another example of an algorithm on graphs of bounded pathwidth, we show how to count the number of perfect matchings in a graph of bounded pathwidth. Note that the problem COUNTING PERFECT MATCHINGS is #P-complete.

Lemma 5.5. *Let $G = (V,E)$ be a graph on n vertices with a given path decomposition of width at most k. Then the problem* COUNTING PERFECT MATCHINGS *can be solved in time $\mathcal{O}(2^k \cdot k \cdot n)$.*

Proof. As in Lemma 5.4, we construct a nice path decomposition

$$P = (X_1, X_2, \ldots, X_r)$$

of G of width k and for $i \in \{1, 2, \ldots, r\}$, we put

$$V_i = \bigcup_{j=1}^{i} X_j.$$

For every partition (A,B) of X_i, we define $m_i(A,B)$ to be equal to the number of matchings M in $G[V_i]$ such that every vertex of $V_i \setminus B$ is an endpoint of some edge of M, and none of the vertices of B is an endpoint of M. Let us note that the number of perfect matchings in G is equal to $m_r(X_r, \emptyset)$.

For $i = 1$ the computation is trivial, $m_1(X_1, \emptyset) = m_1(\emptyset, X_1) = 0$. For $i > 1$, we compute the values of all m_i by making use of the following recursive formulas. Two cases are possible.

— $X_i = X_{i-1} \cup \{v\}$, for some $v \in V$. Then for every partition (A, B) of X_i, if $v \in A$, we have that

$$m_i(A, B) = \sum_{\substack{u \in A \\ \{u,v\} \in E}} m_{i-1}(A \setminus \{u, v\}, B \cup \{u\}),$$

and

$$m_i(A, B) = m_{i-1}(A, B \setminus \{v\}),$$

if $v \notin A$.
— $X_i = X_{i-1} \setminus \{v\}$, for some $v \in V$. In this case, for every partition (A, B) of X_i, we have

$$m_i(A, B) = \{m_{i-1}(A \cup \{v\}, B).$$

By making use of the recurrences, the computations of $m_i(A, B)$ for each i can be done in time $\mathcal{O}(2^{|X_i|}|X_i|)$ and by making use of space $\mathcal{O}(2^{|X_i|})$. Thus the total running time is

$$\mathcal{O}(\sum_{i=1}^{r} 2^{|X_i|}|X_i|) = \mathcal{O}(2^k \cdot k \cdot n).$$

$\square$

Exercise 5.6. Prove that the perfect matchings in an n-vertex graph with an independent set of size i can be counted in time $\mathcal{O}^*(2^{n-i})$. In particular, for bipartite graphs in time $\mathcal{O}^*(2^{n/2})$. Hint: Prove that the pathwidth of such a graph is at most $n - i$.

Another example is the MINIMUM BISECTION problem. In the MINIMUM BISECTION problem, one is asked to partition the vertex set of a graph $G = (V, E)$ into two sets of almost equal size, i.e. the sets are of size $\lceil |V|/2 \rceil$ and $\lfloor |V|/2 \rfloor$, such that the number of edges between the sets is minimized.

Exercise 5.7. Let $G = (V, E)$ be a graph on n vertices with a given path decomposition of width at most k. Prove that the following algorithms can be established.

- The MINIMUM BISECTION problem is solvable in time $\mathcal{O}(2^k \cdot k \cdot n^3)$.
- The MAXIMUM INDEPENDENT SET problem is solvable in time $\mathcal{O}(2^k \cdot k \cdot n)$.
- The MINIMUM DOMINATING SET problem is solvable in time $\mathcal{O}(3^k \cdot k \cdot n)$.

5.2 Graphs of Maximum Degree 3

In this section we show how dynamic programming on graphs of small pathwidth (or treewidth) combined with combinatorial bounds on the pathwidth can be used to design exact algorithms. The main combinatorial bound we prove in this section is that the pathwidth of an n-vertex graph with all vertices of degree at most 3, is roughly at most $n/6$.

We need the following result.

Lemma 5.8. *For any tree T on $n \geq 3$ vertices, $\mathrm{pw}(T) \leq \log_3 n$.*

Exercise 5.9. Prove Lemma 5.8.

The following deep result of Monien and Preis [158] is crucial for this section.

Theorem 5.10. *For any $\varepsilon > 0$, there exists an integer $n_\varepsilon < (4/\varepsilon) \cdot \ln(1/\varepsilon) \cdot (1 + 1/\varepsilon^2)$ such that for every graph $G = (V, E)$ of maximum degree at most 3 satisfying $|V| > n_\varepsilon$, there is a partition V_1 and V_2 of V with $\left| |V_1| - |V_2| \right| \leq 1$ such that*

$$\mathrm{CUT}(V_1, V_2) \leq (1/6 + \varepsilon)|V|.$$

The following technical lemma says that for any vertex subset X of a graph of maximum degree 3, it is possible to construct a path decomposition with the last bag exactly X, and of width roughly at most $\max\{|X|, \lfloor n/3 \rfloor + 1\}$.

Lemma 5.11. *Let $G = (V, E)$ be a graph on n vertices and with maximum degree at most 3. Then for any vertex subset $X \subseteq V$ there is a path decomposition*

$$P = (X_1, X_2, \ldots, X_r)$$

of G of width at most

$$\max\{|X|, \lfloor n/3 \rfloor + 1\} + (2/3) \log_3 n + 1$$

and such that $X = X_r$.

Proof. We prove the lemma by induction on the number of vertices in a graph. For a graph on one vertex the lemma is trivial. Suppose that the lemma holds for all graphs on fewer than n vertices for some $n > 1$.

Let $G = (V, E)$ be a graph on n vertices and let $X \subseteq V$. Different cases are possible.

Case 1. There is a vertex $v \in X$ such that $N(v) \setminus X = \emptyset$, *i.e. v has no neighbors outside X.* By the induction assumption, there is a path decomposition $(X_1, X_2, \ldots, X_r)$ of $G \setminus \{v\}$ of width at most

$$\max\{|X| - 1, \lfloor (n-1)/3 \rfloor\} + (2/3) \log_3 (n-1) + 1$$

and such that $X \setminus \{v\} = X_r$. By adding v to the bag X_r we obtain a path decomposition of G of width at most $\max\{|X|, \lfloor n/3 \rfloor\} + (2/3) \log_3 n + 1$.

Case 2. There is a vertex $v \in X$ such that $|N(v) \setminus X| = 1$, *i.e. v has exactly one neighbor outside X.* Let u be such a neighbor. By the induction assumption for

$G \setminus \{v\}$ and for $X \setminus \{v\} \cup \{u\}$, there is a path decomposition $P' = (X_1, X_2, \ldots, X_r)$ of $G \setminus \{v\}$ of width at most

$$\max\{|X|, \lfloor (n-1)/3 \rfloor\} + (2/3)\log_3 (n-1) + 1$$

and such that $X \setminus \{v\} \cup \{u\} = X_r$. We create a new path decomposition P from P' by adding bags $X_{r+1} = X \cup \{u\}$, $X_{r+2} = X$, i.e.

$$P = (X_1, X_2, \ldots, X_r, X_{r+1}, X_{r+2}).$$

The width of this decomposition is at most

$$\max\{|X|, \lfloor n/3 \rfloor\} + (2/3)\log_3 n + 1.$$

Case 3. For every vertex $v \in X$, $|N(v) \setminus X| \geq 2$. We consider two subcases.

Subcase 3.A. $|X| \geq \lfloor n/3 \rfloor + 1$. The number of vertices in $G \setminus X$ is $n - |X|$. The number of edges in $G \setminus X$ is at most

$$\frac{3(n - |X|) - 2|X|}{2} = \frac{3n - 5|X|}{2} = n - |X| + \frac{n - 3|X|}{2}$$

$$\leq n - |X| + \frac{n - 3(\lfloor n/3 \rfloor + 1)}{2}$$

$$< n - |X| + \frac{n - 3(n/3 - 1) + 3}{2}$$

$$= n - |X| = |V \setminus X|$$

Since the number of edges in the graph $G \setminus X$ is less than $|V \setminus X|$, we have that there is a connected component $T = (V_T, E_T)$ of $G \setminus X$ that is a tree. Note that $|V_T| \leq (2n)/3$. It is not difficult to prove by making use of induction on the height of the tree T (see Exercise 5.8), that there is a path decomposition $P^1 = (X_1, X_2, \ldots, X_r)$ of T of width at most $(2/3)\log_3 n$. By the induction assumption, there is a path decomposition $P^2 = (Y_1, Y_2, \ldots, Y_t = X)$ of $G \setminus V_T$ of width at most $|X| + 1$. The desired path decomposition P of width $\leq |X| + (2/3)\log_3 n + 1$ is formed by adding $X = Y_t$ to all bags of P^1 and appending the altered P^1 to P^2. In other words,

$$P = (Y_1, Y_2, \ldots, Y_t, X_1 \cup X, X_2 \cup X, \ldots, X_r \cup X, X).$$

Case 3.B. $|X| \leq \lfloor n/3 \rfloor$, i.e. every vertex $v \in X$ has at least two neighbors outside X. In this case we choose a set $S \subseteq V \setminus X$ of size $\lfloor n/3 \rfloor - |X| + 1$. If there is a vertex of $X \cup S$ having at most one neighbor in $V \setminus (X \cup S)$, we are in Case 1 or in Case 2. If every vertex of $X \cup S$ has at least two neighbors in $V \setminus (X \cup S)$, then we are in Case 3.A. For each of these cases, there is a path decomposition $P = (X_1, X_2, \ldots, X_r)$ of width $\leq \lfloor n/3 \rfloor + 2$ such that $X_r = X \cup S$. By adding bag $X_{r+1} = X$, we obtain a path decomposition of width at most $\lfloor n/3 \rfloor + 2$.   □

Now we are ready to prove one of the main theorems of this section.

Theorem 5.12. *For any $\varepsilon > 0$, there exists an integer n_ε such that for every graph $G = (V,E)$ with maximum degree at most three and with $|V| > n_\varepsilon$, $\mathrm{pw}(G) \le (1/6 + \varepsilon)|V|$.*

Proof. For $\varepsilon > 0$, let G be a graph on $n > n_\varepsilon \cdot (8/\varepsilon) \cdot \ln(1/\varepsilon) \cdot (1 + 1/\varepsilon^2)$ vertices, where n_ε is given in Theorem 5.10, such that G has maximum degree at most three. By Theorem 5.10, there is a partition of V into parts V_1, V_2 of sizes $\lfloor n/2 \rfloor$ and $\lceil n/2 \rceil$ such that there are at most $(\frac{1}{6} + \frac{\varepsilon}{2})|V|$ edges with endpoints in V_1 and V_2. Let $\partial(V_1)$ ($\partial(V_2)$) be the set of vertices in V_1 (V_2) having a neighbor in V_2 (V_1). Note that $|\partial(V_i)| \le (1/6 + \frac{\varepsilon}{2})n$, $i = 1, 2$.

By Lemma 5.11, there is a path decomposition $P_1 = (A_1, A_2, \ldots, A_p)$ of $G[V_1]$ and a path decomposition $P_3 = (C_1, C_2, \ldots, C_s)$ of $G[V_2]$ of width at most

$$\max\{(1/6 + \frac{\varepsilon}{2})n, \lfloor n/6 \rfloor + 1\} + (2/3)\log_3 n + 1 \le (1/6 + \varepsilon)n$$

such that $A_p = \partial(V_1)$ and $C_1 = \partial(V_2)$.

It remains to show how the path decomposition P of G can be obtained from path decompositions P_1 and P_3. To construct P we show that there is a path decomposition $P_2 = (B_1, B_2, \ldots, B_r)$ of $G[\partial(V_1) \cup \partial(V_2)]$ of width $\le (1/6 + \varepsilon)n$ and with $B_1 = \partial(V_1)$, $B_r = \partial(V_2)$. The union of P_1, P_2, and P_3, i.e. $(A_1, \ldots, A_p = B_1, \ldots, B_r = C_1, \ldots, C_s)$ will be a path decomposition of G of width at most $(1/6 + \varepsilon)n$.

The path decomposition $P_2 = (B_1, B_2, \ldots, B_r)$ is constructed as follows. We put $B_1 = \partial(V_1)$. In a bag B_j, where $j \ge 1$ is odd, we choose a vertex $v \in B_j \setminus \partial(V_2)$. We put $B_{j+1} = B_j \cup N(v) \cap \partial(V_2)$ and $B_{j+2} = B_{j+1} \setminus \{v\}$. Since we always remove a vertex of $\partial(V_1)$ from B_j (for odd j), we arrive finally at the situation when a bag B_r contains only vertices of $\partial(V_2)$.

To conclude the proof, we argue that for any $j \in \{1, 2, \ldots, k\}$, $|B_j| \le (1/6 + \varepsilon)n + 1$. Let D_m, $m = 1, 2, 3$, be the set of vertices in $\partial(V_1)$ having exactly m neighbors in $\partial(V_2)$. Thus

$$|B_1| = |\partial(V_1)| = |D_1| + |D_2| + |D_3|$$

and

$$|D_1| + 2 \cdot |D_2| + 3 \cdot |D_3| \le (1/6 + \varepsilon)n.$$

Therefore,

$$|B_1| \le (1/6 + \varepsilon)n - |D_2| - 2 \cdot |D_3|.$$

For a set B_j, $j \in \{1, 2, \ldots, k\}$, let $D_2' = B_j \cap D_2$ and $D_3' = B_j \cap D_3$. Every time ℓ ($\ell \le 3$) vertices are added to a bag, one vertex is removed from the next bag. Thus

$$
\begin{aligned}
|B_j| &\le |B_1| + |D_2 \setminus D_2'| + 2 \cdot |D_3 \setminus D_3'| + 1 \\
&\le (1/6 + \varepsilon)n - (|D_2| - |D_2 \setminus D_2'|) - 2 \cdot (|D_3| - |D_3 \setminus D_3'|) + 1 \\
&\le (1/6 + \varepsilon)n + 1.
\end{aligned}
$$

$\square$

The proof of Theorem 5.10 is constructive and can be turned into a polynomial time algorithm constructing for any large graph G of maximum degree at most three a cut (V_1, V_2) of size at most $(1/6 + \varepsilon)|V|$ and such that $||V_1| - |V_2|| \leq 1$. The proof of Theorem 5.12 is also constructive and can be turned into a polynomial time algorithm constructing a path decomposition of graphs G of maximum degree 3 of width $\leq (1/6 + \varepsilon)|V|$.

Theorem 5.12 can be used to obtain algorithms on graphs with maximum degree 3. For example, combining it with dynamic programming algorithms for graphs of bounded pathwidth (see Lemma 5.4 and Exercise 5.7), we establish the following corollary.

Corollary 5.13. *For every $\varepsilon > 0$, on n-vertex graphs with maximum degree 3:*

- MAXIMUM CUT, MINIMUM BISECTION *and* MAXIMUM INDEPENDENT SET *are solvable in time* $\mathcal{O}^*(2^{(1/6+\varepsilon)n})$;
- MINIMUM DOMINATING SET *is solvable in time* $\mathcal{O}^*(3^{(1/6+\varepsilon)n})$.

It is possible to extend the upper bound on the pathwidth of large graphs of maximum degree 3, given in Theorem 5.12, in the following way.

Theorem 5.14. *For any $\varepsilon > 0$, there exists an integer n_ε such that for every graph G with $n > n_\varepsilon$ vertices,*

$$\mathrm{pw}(G) \leq \frac{1}{6}n_3 + \frac{1}{3}n_4 + \frac{13}{30}n_5 + n_{\geq 6} + \varepsilon n,$$

where n_i is the number of vertices of degree i in G for any $i \in \{3, \ldots, 5\}$ and $n_{\geq 6}$ is the number of vertices of degree at least 6.

Theorem 5.14 can be proved by induction on the number of vertices of a given degree. By making use of Theorem 5.14, it is possible to bound the pathwidth of a graph in terms of m, the number of edges.

Lemma 5.15. *For any $\varepsilon > 0$, there exists an integer n_ε such that for every graph G with $n > n_\varepsilon$ vertices and m edges,*

$$\mathrm{pw}(G) \leq 13m/75 + \varepsilon n.$$

Proof. First, suppose that G does not have vertices of degree larger than 5. Then every edge in G contributes at most

$$\max_{3 \leq d \leq 5} \left\{ \frac{2\beta_d}{d} \right\}$$

to the pathwidth of G, where $\beta_3 = 1/6, \beta_4 = 1/3, \beta_5 = 13/30$ are the values from Theorem 5.14. The maximum is obtained for $d = 5$ and it is $13/75$. Thus, the result follows for graphs of maximum degree at most 5.

Finally, if G has a vertex v of degree at least 6, then we use induction on the number of vertices of degree at least 6. The base case has already been proved and the inductive argument is as follows:

$$\mathrm{pw}(G) \leq \mathrm{pw}(G \setminus v) + 1 \leq 13(m-6)/75 + 1 < 13m/75.$$

$\square$

As a consequence of this bound, for graphs with at most $75n/13$ edges, for various NP-hard problems, such as MAXIMUM CUT, the dynamic programming algorithm for graphs of bounded pathwidth has running time $\mathcal{O}^*(c^n)$ for some $c < 2$.

5.3 Counting Homomorphisms

In Chap. 4 we used graph homomorphisms to count isomorphic subgraphs. However counting the homomorphisms from a graph F to a graph G is a difficult problem on its own and thus this idea worked fine only for graphs F like cycles or cliques. In this section we show how this idea can be pushed forward by counting the homomorphisms from a graph F of bounded treewidth to a graph G. As an example of this approach, we show how to compute the bandwidth of a tree in time $\mathcal{O}^*(2^n)$.

In Sect. 5.1 the notion of a nice path decomposition was defined. Such path decompositions were useful in performing dynamic programming. Now we need a similar notion for treewidth.

A tree decomposition $(\{X_i : i \in I\}, T = (I, F))$ of a graph $G = (V, E)$ is *rooted* if the tree is rooted. Since in a rooted tree all nodes can be ordered corresponding to their distance from the root, we can also speak of children and parents of bags in rooted tree decompositions.

A rooted tree decomposition $(\{X_i : i \in I\}, T = (I, F))$ of G is *nice* if

- Every node $i \in I$ of T has at most two children;
- If a node i has two children j and k, then $X_i = X_j = X_k$ (such a node is called a *join* node);
- If a node i has one child j, then

  - either $X_i = X_j \cup \{v\}$, for some $v \in V$ (in this case i is called an *introduce* node);

  - or $X_i = X_j \setminus \{v\}$, for some $v \in V$ (in this case i is called a *forget* node);

- Every leaf node of T contains exactly one vertex of G.

A nice tree decomposition of the graph from Fig. 5.1 is given in Fig. 5.3.
We refer to the book of Kloks [132] for the proof of the following Lemma.

Lemma 5.16. *For any integer $k \geq 1$, given a tree decomposition of a graph G of width k it is possible to transform it in time $\mathcal{O}(n)$ into a nice tree decomposition of G of width k and with at most $4n$ nodes.*

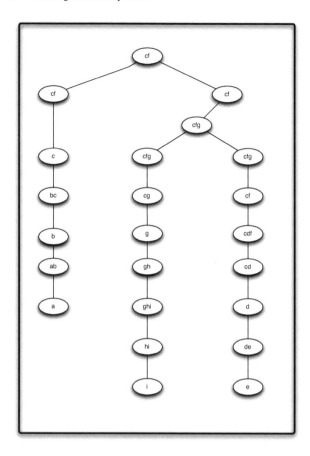

Fig. 5.3 Nice tree decomposition of the graph from Fig. 5.1

Now we are ready to show how homomorphisms from a graph of bounded treewidth can be counted.

Lemma 5.17. *Let F be a graph on n_F vertices given together with its nice tree decomposition of width t. Then the number of homomorphisms $\hom(F, G)$ from F to an n-vertex graph G can be computed in time $\mathcal{O}(n_F \cdot n^{t+1} \cdot \min\{t, n\})$ and space $\mathcal{O}(n_F \cdot n^{t+1})$.*

Proof. Let $(\{X_i \mid i \in I\}, T = (I, F))$ be a rooted tree decomposition of width t with root node r of the graph $F = (V_F, E_F)$. For a node $i \in I$ we define the set V_i as the set of all vertices of V_F that are contained in bag X_i or in some bag corresponding to a node j which is below i in T. In other words,

$$V_i = \bigcup_j X_j,$$

where j runs through all nodes of T such that i is on the path from r to j. We also define $F_i = F[V_i]$, the subgraph of F induced by V_i. For every node i and mapping $\phi : X_i \rightarrow V_G$, we show how to compute the value $\hom(F_i, G, \phi)$, which is the number of homomorphisms f from F_i to G such that for every vertex $v \in X_i$, $f(v) = \phi(v)$. Since $F_r = F$, we have that

$$\hom(F, G) = \sum_{\phi : X_r \rightarrow V_G} \hom(F_r, G, \phi).$$

We compute the values of $\hom(F_i, G, \phi)$ by performing dynamic programming. We start from leaves of T. Every leaf i of T consists of one vertex of F and in this case for every mapping $\phi : X_i \rightarrow V_G$ we put $\hom(F_i, G, \phi) = 1$.

If i is an introduce node of T with a child j, then we compute $\hom(F_i, G, \phi)$ as follows. Let $v = X_i \setminus X_j$. Then the graph F_i is obtained from F_j by adding vertex v and some edges with endpoints in v. Let $N_i(v) = N_{F_i}(v)$ be the set of neighbors of v in F_i. Then by the properties of tree decompositions (Exercise 5.2), we have that

$$N_i(v) = N_{F_i}(v) \cap X_i,$$

i.e. all neighbors of v in F_i are in X_i. Because of this, every homomorphism $\phi : F_i \rightarrow G$ is also a homomorphism from F_j to G such that the neighbors of v in F_i are mapped to neighbors of $\phi(v)$ in G. Therefore,

$$\hom(F_i, G, \phi) = \sum \hom(F_j, G, \psi),$$

where the sum is taken over all mappings $\psi : X_j \rightarrow V(G)$ such that $\psi(N_i(v)) \subseteq N_G(v)$. For every mapping $\psi : X_j \rightarrow V_G$, there are at most $\min\{|X_i|, n\}$ mappings $\phi : X_i \rightarrow V_G$ "extending" ψ, and thus the values $\hom(F_i, G, \phi)$ for all mappings $\phi : X_i \rightarrow V_G$ can be computed in time $\mathcal{O}(|X_i|^n \cdot \min\{|X_i|, n\})$.

If i is a join node with children j and k, then by the properties of tree decomposition, there are no edges connecting vertices of $F_j \setminus X_i$ with vertices $F_k \setminus X_i$. Then for every mapping $\phi : X_i \rightarrow V_G$, we have that

$$\hom(F_i, G, \phi) = \hom(F_j, G, \phi) \cdot \hom(F_k, G, \phi).$$

Thus for a join node the values of $\hom(F_i, G, \phi)$ for all mappings ϕ are computable in time $\mathcal{O}(|X_i|^n)$.

If i is a forget node with a child j, then

$$\hom(F_i, G, \phi) = \hom(F_j, G, \phi).$$

The number of nodes in T is $\mathcal{O}(n_F)$, and thus the number of operations required to compute $\hom(F, G)$ is $\mathcal{O}(n_F \cdot n^{t+1} \cdot \min\{t, n\})$. For every node we keep the number of homomorphisms corresponding to every mapping ϕ, and thus the space used by the algorithm is $\mathcal{O}(n_F \cdot n^{t+1})$. $\qquad\square$

Lemma 5.17 combined with Theorem 4.18 yields the following result.

Theorem 5.18. *Let F be a graph with n_F vertices and G be a graph with n vertices such that $n_F \leq n$, given together with a tree-decomposition of width t of F. Then $\text{sub}(F,G)$ can be computed in time*

$$\mathcal{O}(\sum_{i=0}^{n_F} \binom{n}{i} \cdot n_F \cdot n^{t+1} \cdot t)$$

and space $\mathcal{O}(n_F \cdot n^{t+1})$.

Proof. Observe that $\text{aut}(F) = \text{inj}(F,F)$. Hence, using Theorem 4.18 together with Lemma 5.17 we can compute $\text{aut}(F)$ in time $\mathcal{O}(2^{n_F} \cdot n_F^{t+2} \cdot t)$ and space $\mathcal{O}(\log n_F \cdot n_F^{t+1})$.

Now we use Theorem 4.18 and Lemma 5.17 to compute the value of $\text{inj}(F,G)$ in time

$$\mathcal{O}(\sum_{i=0}^{n_F} \binom{n}{i} \cdot n_F \cdot n^{t+1} \cdot t)$$

and space $\mathcal{O}(\log n_F \cdot n^{t+1})$. We also know (Proposition 4.16), that

$$\text{sub}(F,G) = \frac{\text{inj}(F,G)}{\text{aut}(F)}$$

which allows us to conclude the proof of the theorem.   □

Let us consider some examples showing how to use Theorem 5.18.

Bandwidth. In the BANDWIDTH MINIMIZATION problem we are given an undirected graph $G = (V,E)$ on n vertices. The task is to compute the *bandwidth* of the graph G. This means we wish to compute a linear ordering of the vertices such that the maximum stretch of any edge of G in the ordering is minimized.

The problem has its origin in computations with sparse matrices (this is also the origin of the name of the problem). Given a sparse matrix, i.e. a matrix with many zero entries, the task is to permute the rows and columns such that all non-zero entries appear in a narrow band close to the main diagonal.

Using graphs this problem can be formally stated as follows. A *layout* of a graph $G = (V,E)$ on n vertices is a bijection $f : V \rightarrow \{1,\dots,n\}$. The *bandwidth* of G is

$$\min\{\max\{|f(u) - f(v)| : \{u,v\} \in E\}\},$$

where the minimum is taken over all layouts of f of G. The objective of the BANDWIDTH MINIMIZATION problem is to find an optimal layout, i.e. layout $f : V \rightarrow \{1,\dots,n\}$ such that $\max_{\{u,v\} \in E} |f(u) - f(v)|$ is minimized.

The BANDWIDTH MINIMIZATION problem can be seen as a SUBGRAPH ISO-MORPHISM problem. Let P_n be a path on n vertices. The rth power P_n^r of P_n is the graph obtained from P_n by adding edges between any pair of distinct vertices of the P_n being in distance at most r in the graph P_n. For an example, see Fig. 5.4.

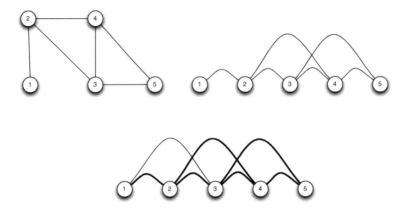

Fig. 5.4 A graph G, its layout of bandwidth 2, and an embedding of G into P_5^2

Lemma 5.19. *Let G be a graph on n vertices. Then G has a layout of bandwidth b if and only if there is an injective homomorphism from G to P_n^b.*

Proof. Let f be a function which gives a layout of bandwidth b for $G = (V,E)$. Notice that f can be also viewed as a mapping from G to P_n. Now, if $\{u,v\} \in E$, then we add an edge $\{f(u), f(v)\}$ to P_n. Let the resulting graph be P_n'. Notice that no edge has stretched to more than length b in G which in turn implies that G is a subgraph of $P_n' \subseteq P_n^b$. The other direction follows similarly.   □

Consequently the BANDWIDTH MINIMIZATION problem is a special case of the SUBGRAPH ISOMORPHISM problem, and by Theorem 5.18, we have the following corollary.

Corollary 5.20. *Given a graph G on n vertices together with a tree decomposition of width t, it is possible to compute the number of its minimum bandwidth layouts in time $2^{n+t\log_2 n} n^{\mathcal{O}(1)}$. In particular, if G is a tree then we can compute the number of its minimum bandwidth layouts in time $\mathcal{O}(2^n \cdot n^3)$ and polynomial space.*

Packing Problems. Let $\mathcal{H}$ be a class of graphs. In the PACKING problem we are given an undirected graph $G = (V,E)$. The task is to find the maximum number of vertex-disjoint copies of graphs $H \in \mathcal{H}$ in the graph $G = (V,E)$. The *packing number* of G with respect to the set $\mathcal{H}$ denoted by $\text{pack}_{\mathcal{H}}(G)$ is the maximum integer k for which there is a partition $V_1, V_2, \ldots, V_k$ of V such that every graph $G[V_i]$ contains at least one graph H of $\mathcal{H}$ as a subgraph.

In the following corollary we give a result on cycle packing.

Theorem 5.21. *Let $G = (V,E)$ be a graph on n vertices and $\mathcal{H}$ be the class of graphs consisting of cycles. Then* **$\text{pack}_{\mathcal{H}}(G)$** *can be found in time $2^{n+\mathcal{O}(\sqrt{n})}$ and polynomial space.*

Proof. Given a graph class $\mathcal{H}$ and the input graph G, to solve **pack**$_{\mathcal{H}}(G)$, for every $\ell \leq n$, we take all possible unordered partitions of ℓ. By Theorem 4.21, the number of such partitions is $p(\ell) = 2^{\mathcal{O}(\sqrt{\ell})}$. Furthermore as already mentioned in Sect. 4.4, all these partitions can be listed in time $2^{\mathcal{O}(\sqrt{\ell})}$. For each such partition $p = (n_1, n_2, \ldots, n_k)$, we construct a graph F_p consisting of disjoint copies of cycles on n_i vertices, $1 \leq i \leq k$. Each of these graphs F_p is of treewidth at most 2, and now the result follows from Theorem 5.18. $\square$

Finally, we would like to mention that this result can easily be generalized to the case where $\mathcal{H}$ is a class of graphs of bounded treewidth.

5.4 Computing Treewidth

In this section we describe an exact algorithm computing the treewidth of a graph. The algorithm is strongly based on the theory of potential maximal cliques developed by Bouchitté and Todinca.

There are several equivalent definitions of treewidth, and for this algorithm we use the definition of treewidth in terms of chordal graphs. A graph H is *chordal* (or *triangulated*) if every cycle of length at least four has a chord, i.e. an edge between two non-consecutive vertices of the cycle. A *triangulation* of a graph $G = (V, E)$ is a chordal graph $H = (V, E')$ such that $E \subseteq E'$. Triangulation H is a *minimal triangulation* if for any set E'' with $E \subseteq E'' \subset E'$, the graph $F = (V, E'')$ is not chordal. See Fig. 5.5 for an example of a minimal triangulation of a graph.

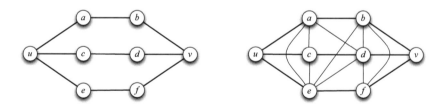

Fig. 5.5 A Graph G and a minimal triangulation of G

Let $\omega(H)$ be the maximum clique-size of a graph H.

Theorem 5.22 (folklore). *For any graph G, $\text{tw}(G) \leq k$ if and only if there is a triangulation H of G such that $\omega(H) \leq k+1$.*

Minimal Separator. Let u and v be two non adjacent vertices of a graph $G = (V, E)$. A set of vertices $S \subseteq V$ is a *u, v-separator* if u and v are in different connected components of the graph $G \setminus S$. S is a *minimal u, v-separator* of G if no proper subset of S is a u, v-separator. For example, in graph G in Fig. 5.5, the following sets

are minimal u,v-separators: $\{a,c,e\}, \{a,c,f\}, \{a,d,e\}, \{a,d,f\}, \{b,c,e\}, \{b,c,f\},$
$\{b,d,e\}, \{b,d,f\}$. We say that S is a *minimal separator* of G if there are two vertices
u and v such that S is a minimal u,v-separator. Notice that a minimal separator can
be a subset of another minimal separator separating a different pair of vertices. We
denote by Δ_G the set of all minimal separators of G.

Potential Maximal Clique. A set of vertices $\Omega \subseteq V$ of a graph G is called a *potential
maximal clique* if there is a minimal triangulation H of G such that Ω is a maximal
clique of H. For instance, the sets $\{a,c,e,b,d\}$ and $\{u,a,c,e\}$ of the graph G in
Fig. 5.5 are some of the potential maximal cliques of G. We denote by Π_G the set of
all potential maximal cliques of G.

The algorithm computing the treewidth of a graph follows from the following
two theorems.

The first theorem reduces the problem of computing the treewidth of a graph to
the problem of enumerating its potential maximal cliques and minimal separators.

Theorem 5.23. *There is an algorithm that, given an n-vertex graph G together with
the list of its minimal separators Δ_G and the list of its potential maximal cliques
Π_G, computes the treewidth of G in $\mathcal{O}^*(|\Pi_G| + |\Delta_G|)$ time. Moreover, the optimal
tree decomposition can be constructed within the same running time.*

The second theorem is used to bound and enumerate the potential maximal
cliques and minimal separators.

Theorem 5.24. *For any n-vertex graph G, $|\Delta_G| = \mathcal{O}(1.6181^n)$ and $|\Pi_G| = \mathcal{O}(1.7549^n)$.
Moreover, it is possible to enumerate all minimal separators in time $\mathcal{O}(1.6181^n)$ and
all potential maximal cliques in time $\mathcal{O}(1.7549^n)$.*

Combining Theorems 5.23 and 5.24, we obtain the following corollary.

Corollary 5.25. *The treewidth of an n-vertex graph is computable in time $\mathcal{O}(1.7549^n)$.*

Proofs of Theorems 5.23 and 5.24 are based on deep and technical combinatorial results. In the following two subsections we only sketch the proofs of these
theorems.

5.4.1 Computing the Treewidth Using Potential Maximal Cliques

In this subsection we prove Theorem 5.23. We assume that we are given a set of all
potential maximal cliques Π_G and minimal separators Δ_G. The algorithm computing
the treewidth of G from Π_G and Δ_G is based on the following relation of treewidth
and pursuit-evasion games played on a graph.

As we already have seen, there are several parameters equivalent to the treewidth
of a graph. There is one more parameter, this time defined in terms of a Cops and
Robber game. The game is played on a graph $G = (V,E)$ by two players, the *Cop-
player* and the *Robber-player*, each having her *pawns* (cops or a robber, respectively) on the vertices of G. Cops and the robber occupy vertices of the graphs. The

goal of the Cop-player is to catch the robber. The robber is caught when a cop is placed on the vertex currently occupied by the robber. The robber always knows where cops are, and moves arbitrarily fast from vertex to vertex via edges of the graph, but it cannot pass a vertex occupied by a cop without being caught. The cops are moved from vertex to vertex, say, flying on helicopters.

More formally, let $G = (V, E)$ be a graph, where k cops are trying to catch the robber. Every step of the game is characterized by the position (C, R), where $C, R \subseteq V$, $|C| \leq k$, and R is one of the connected components of $G \setminus C$. Vertex set C is the set of vertices occupied by cops, and R tells where the robber is located— since it can move arbitrarily fast, all that matters is the connected component containing it.

The game starts with the initial position $(\emptyset, V)$, there are no cops on the graph and the robber can be anywhere. The game proceeds in steps. Suppose that at the ith step, the position of the players is (C, R). Now the cops select a subset C' such that $C' \supset C$ (some of the cops are placed on the graph) or $C' \subset C$ (some of the cops are removed from the graph). The robber's actions are

- If $C' \subset C$, i.e. this is a removal step, then the robber occupies the connected component R' of $G \setminus C'$ such that $R \subseteq R'$. The game proceeds to the $(i+1)$th step with position (C', R');
- If $C' \supset C$, i.e. this is a placement step when some cops are placed on the graph, then the robber selects a non-empty connected component R' of $G \setminus C'$ such that $R' \subseteq R$, and the game proceeds to the $(i+1)$th step with position (C', R'). If there is no such non-empty component, i.e. $C' \supseteq R$, then the robber has no place to run and the Cop-player wins.

We say that k cops have a winning strategy on G, if for any strategy of the Robber-player, the Cop-player can win with k cops in a finite number of steps. For example, on any tree, 2 cops have a winning strategy: One of the cops selects a vertex v of tree T, and then the robber can only be in one of the connected components C of $T \setminus \{v\}$. Then the second cop steps on vertex the $u \in C$ adjacent to v, thus reducing the space, i.e. available vertices, of the robber. Thus in at most n steps the cops can catch the robber. This strategy can be easily generalised to show that on graphs of treewidth k, $k+1$ cops have a winning strategy. The non-trivial and deep result is the converse statement, which is due to Seymour and Thomas [204].

Theorem 5.26. *For every $k \geq 0$, the following are equivalent*

- *$k+1$ cops have a winning strategy on G;*
- *The treewidth of G is at most k.*

Moreover, Seymour and Thomas prove that if k cops can win, they can always do it in a monotone way, which means that for each step from position (C, R) to (C', R'), $R' \subset R$ if this is a placement step, and $R' = R$ if this is a removal step. Thus the cops can always win by making at most $2n - 1$ steps.

It is possible to prove that if $k+1$ cops have a winning strategy on G, then $k+1$ cops can win even when their moves are constrained by the following conditions:

after every placement step, the set of vertices C occupied by cops is one of the potential maximal cliques of G, and after a removal step, the set of vertices occupied by cops is a minimal separator. Indeed, by Theorem 5.26, if $k+1$ cops have a winning strategy, then the treewidth of G is at most k.

By Theorem 5.22, there is a minimal triangulation H of G such that the size of a maximal clique of H is at most $k+1$. We leave it as an exercise to prove that $k+1$ cops on H can win in such a way that at every step the vertex set C occupied by the cops is either a maximal clique or a minimal separator of H. (Hint: Make use of the fact that every chordal graph has a tree decomposition such that every bag of the decomposition is a maximal clique.) Such a strategy is very similar to the strategy of 2 cops on a tree; instead of two adjacent vertices, which form a maximal clique in a tree, cops now occupy a maximal clique and instead of one vertex they occupy a minimal separator. Every maximal clique of H is a potential maximal clique of G and every minimal separator of H is also a minimal separator of G. We leave the proof of this fact as another exercise.

Now for every $k \geq 1$, we want to find out whether k cops have a winning strategy on on a graph $G = (V, E)$, which by Theorem 5.26, implies that the treewidth of G is at most $k-1$. To solve the Cops and Robber problem, we use a standard technique from games on graphs. The idea is to represent all configurations of the game and their dependencies by a directed graph called the *arena*. The vertex set of the arena is partitioned into two sets of nodes V_1 and V_2. It also also has a specified node $v \in V_1$ corresponding to the initial position of the game and a node $u \in V_2$ corresponding to the final positions. There are two players, Player 1 and Player 2, who alternatively move a token along arcs of the arena. The game starts by placing the token on v, and then the game alternates in rounds, starting with Player 1. At each round, if the token is on a vertex from V_i, $i = 1,2$, then the ith player slides the token from its current position along an arc to a new position. Player 1 wins if he manage to slide the token to the final position node u. Otherwise, Player 2 wins.

We construct an auxiliary directed graph H, the arena, as follows. Let Π_G^k and Δ_G^k be the sets of potential maximal cliques and, correspondingly, minimal separators of G of size at most k. The vertex set of arena H consists of all possible pairs (C,R), where C is an element of Π_G^k or Δ_G^k, and R is a connected component of $G \setminus C$. To distinguish the vertices of the arena and the original graph, we will refer to the vertices of the arena as to nodes. The node set V_1 of Player 1 is the set of all nodes corresponding to minimal separators, and the node set V_2 is the set of nodes corresponding to potential maximal cliques.

Exercise 5.27. Prove that no potential maximal clique can be a minimal separator.

Thus every node of H represents a possible situation of the game when cops occupy the set of vertices C, and the robber is in the component R. We also add a node v corresponding to the initial position of the game $(\emptyset, V)$—no cops are on the graph and the robber can choose any vertex of the graph, and node u corresponds to the winning position of the cops. Since the number of connected components for every $C \in \Pi_G^k \cup \Delta_G^k$ is at most $n-1$, we have that the number of nodes in the arena is at most

$$(n-1)|\Pi_G^k| + (n-1)|\Delta_G^k| + 2 \leq n(|\Pi_G| + |\Delta_G|).$$

The arcs of H are of two types: *remove-arcs* and *put-arcs* corresponding to removal and placement of the cops on the graph. Each remove-arc goes from node $(C_1, R_1) \in V_2$ to node $(C_2, R_2) \in V_1$ if $C_1 \in \Pi_G^k$, $C_2 \in \Delta_G^k$, $C_2 \subset C_1$ and $R_1 = R_2$. This arc corresponds to the step of the Cops and Robber game when position (C_2, R_2) is obtained from (C_1, R_1) by removing some cops. Each put-arc goes from node $(C_1, R_1) \in V_1$ to node $(C_2, R_2) \in V_2$ if $C_1 \in \Delta_G^k$, $C_2 \in \Pi_G^k$, $C_1 \subset C_2$, $R_2 \subset R_1$, and $C_2 \subseteq C_1 \cup R_1$. This arc corresponds to the step of the game when position (C_2, R_2) is obtained from (C_1, R_1) by placing cops on C_2 and by the robber selecting one of the components of $G \setminus C_2$. We also add put-arcs from the initial position $v = (\emptyset, V)$ to all nodes of V_2, which means that cops can start by occupying any potential maximal clique. For every node $x = (C, R)$ of V_1, where $C \in \Delta_G^k$ and such that there is a potential maximal clique $\Omega \supseteq C \cup R$ of size at most k, we add a put-arc from x to u. This means that if cops occupy the set C and the robber is in R, then in the next step cops win the game by occupying the vertices of Ω.

For every node of the arena from V_2 (corresponding to a potential maximal clique and one of its components) there is at most one remove-arc going out of it, and the total number of remove-arcs is at most $(n-1)|\Pi_G^k|$. The number of put-arcs going from v is at most $(n-1)|\Pi_G^k|$. For every node $x = (C, R) \in V_1$, the number of put-arcs going from x is at most the number of pairs (C', R'), where $C' \in \Pi_G^k$, $C \subset C' \subseteq C \cup R$, and $R' \subseteq R$. Therefore, the number of put-arcs is upper bounded (up to multiplicative factor n) by the number of the so-called *good* triples, which are the triples of the form (S, Ω, R), where $S \in \Delta_G^k$, $\Omega \in \Pi_G^k$, R is a component of $G \setminus S$, and $R \subset \Omega \subseteq S \cup R$. By making use of some combinatorial properties of potential maximal cliques, it is possible to show that the number of good triples is at most $n|\Pi_G|$. (This part also requires a proof which is not given here.) Thus the number of edges in the arena graph H is $\mathcal{O}(n^2 \cdot |\Pi_G| + |\Delta_G|)$.

By the construction of the arena, we have that Player 1 wins on this arena if and only if k cops have a winning strategy. To find out whether Player 1 wins on arena H, we employ the following simple procedure. We label the final node u of H and proceed further as follows. A node $x \in V_2$ is labelled if every node y of V_1 adjacent to x by arc (x, y) is labelled. In terms of the game it means that to whatever node y Player 2 chooses to shift the token from x, Player 1 is winning. A vertex of $x \in V_1$ is labelled if there is vertex $y \in V_2$ such that (x, y) is arc of the arena. Finally, if as the result of the labelling procedure the starting vertex v is labelled, then Player 1 has a winning strategy. Otherwise, he cannot win. It is easy to implement this labelling procedure in linear (in the size of H) time, which is $\mathcal{O}^*(|\Pi_G| + |\Delta_G|)$.

We have shown that in time $\mathcal{O}^*(|\Pi_G| + |\Delta_G|)$ it is possible to decide whether Player 1 wins on arena H, and thus whether k cops can win on G. By constructing an arena graph for all $k \leq n$, we compute the minimum number of cops having a winning strategy on G, and thus the treewidth of G, in time $\mathcal{O}^*(|\Pi_G| + |\Delta_G|)$.

It is also possible to compute the corresponding winning strategy of the Cops-player and even the corresponding tree decomposition within the same running time.

5.4.2 Counting Minimal separators and Potential Maximal Cliques

We start the proof of Theorem 5.24 with the following combinatorial result concerning the number of *connected vertex subsets*. Note that a vertex subset $B \subseteq V$ is connected if the induced subgraph $G[B]$ is connected.

Lemma 5.28. *Let* $G = (V,E)$ *be a graph. For every* $v \in V$, *and* $b, f \geq 0$, *the number of connected vertex subsets* $B \subseteq V$ *such that*

(i) $v \in B$,
(ii) $|B| = b+1$, *and*
(iii) $|N(B)| = f$

is at most $\binom{b+f}{b}$.

Proof. Let v be a vertex of a graph $G = (V,E)$. For $b + f = 0$, the lemma trivially holds. We proceed by induction assuming that for some $k > 0$ and every b and f such that $b + f \leq k - 1$, the lemma holds. For b and f such that $b + f = k$, we define $\mathcal{B}$ as the set of sets B satisfying $(i), (ii), (iii)$. We claim that

$$|\mathcal{B}| \leq \binom{b+f}{b}.$$

Since the claim always holds for $b = 0$, let us assume that $b > 0$.

Let $N(v) = \{v_1, v_2, \ldots, v_p\}$. For $1 \leq i \leq p$, we define $\mathcal{B}_i$ as the set of all connected subsets B such that

- Vertices $v, v_i \in B$,
- For every $j < i$, $v_j \notin B$,
- $|B| = b+1$,
- $|N(B)| = f$.

Let us note, that every set B satisfying the conditions of the lemma is in some set $\mathcal{B}_i$ for some i, and that for $i \neq j$, $\mathcal{B}_i \cap \mathcal{B}_j = \emptyset$. Therefore,

$$|\mathcal{B}| = \sum_{i=1}^{p} |\mathcal{B}_i|. \tag{5.3}$$

For every $i > f + 1$, $|\mathcal{B}_i| = 0$ (this is because for every $B \in \mathcal{B}_i$, the set $N(B)$ contains vertices $v_1, \ldots, v_{i-1}$ and thus is of size at least $f + 1$.) Thus, (5.3) can be rewritten as follows

$$|\mathcal{B}| = \sum_{i=1}^{f+1} |\mathcal{B}_i|. \tag{5.4}$$

Let G_i be the graph obtained from G by contracting edge $\{v, v_i\}$ (removing the loop, reducing double edges to single edges, and calling the new vertex v) and removing vertices $v_1, \ldots, v_{i-1}$. Then the cardinality of $\mathcal{B}_i$ is equal to the number of the connected vertex subsets B of G_i such that

- $v \in B$,
- $|B| = b$,
- $|N(B)| = f - i + 1$.

By the induction assumption, this number is at most $\binom{f+b-i}{b-1}$ and (5.4) yields that

$$|\mathcal{B}| = \sum_{i=1}^{f+1} |\mathcal{B}_i| \leq \sum_{i=1}^{f+1} \binom{f+b-i}{b-1} = \binom{b+f}{b}.$$

$\square$

The inductive proof of the lemma can easily be turned into a recursive polynomial space enumeration algorithm (we skip the proof here).

Lemma 5.29. *All connected vertex sets of size $b+1$ with f neighbors in a graph G can be enumerated in time $\mathcal{O}(n\binom{b+f}{b})$ by making use of polynomial space.*

The following lemma gives the bound on the number of minimal separators.

Lemma 5.30. *Let Δ_G be the set of all minimal separators in a graph G on n vertices. Then $|\Delta_G| = \mathcal{O}(1.6181^n)$.*

Proof. For $1 \leq i \leq n$, let $f(i)$ be the number of all minimal separators in G of size i. Then

$$|\Delta_G| = \sum_{1}^{n} f(i). \tag{5.5}$$

Let S be a minimal separator of size $\lfloor \alpha n \rfloor$, where $0 < \alpha < 1$. It is an easy exercise to show that for every minimal u, v-separator S, two connected components $u \in C_u$ and $v \in C_v$ of $G[V \setminus S]$ satisfy the property $N(C_1) = N(C_2) = S$. These components are called *full* components associated to S. For every minimal separator there are at least two (but there can be more than two) connected components associated to it. Given a connected vertex set C, there is at most one minimal separator S such that C is a full component associated to S; which requires that $N(C)$ is a minimal separator. Thus instead of counting minimal separators, we count full components.

Let C_1 and C_2 be two full components associated to S. Let us assume that $|C_1| \leq |C_2|$. Then $|C_1| \leq \lfloor (1-\alpha)n/2 \rfloor$. Because C_1 is a full component associated to S, we have that $N(C_1) = S$. Thus, $f(\lfloor \alpha n \rfloor)$ is at most the number of connected vertex sets C of size at most $\lfloor (1-\alpha)n/2 \rfloor$ with neighborhoods of size $|N(C)| = \lfloor \alpha n \rfloor$. Hence, to bound $f(\lfloor \alpha n \rfloor)$ we can use Lemma 5.28 for every vertex of G.

By Lemma 5.28, we have that for every vertex v, the number of full components of size $b+1 = \lfloor (1-\alpha)n/2 \rfloor$ containing v and with neighborhoods of size $\lfloor \alpha n \rfloor$ is at most

$$\binom{b + \lfloor \alpha n \rfloor}{b} \leq \binom{\lfloor (1+\alpha)n/2 \rfloor}{b}.$$

Therefore

$$f(\lfloor \alpha n \rfloor) \leq n \cdot \sum_{i=1}^{\lfloor (1-\alpha)n/2 \rfloor} \binom{i + \lfloor \alpha n \rfloor}{i} < n \cdot \sum_{i=1}^{\lfloor (1-\alpha)n/2 \rfloor} \binom{\lfloor (1+\alpha)n/2 \rfloor}{i}. \quad (5.6)$$

For $\alpha \leq 1/3$, we have

$$\sum_{i=1}^{\lfloor (1-\alpha)n/2 \rfloor} \binom{\lfloor (1+\alpha)n/2 \rfloor}{i} < 2^{\lfloor (1+\alpha)n/2 \rfloor} < 2^{\lfloor 2n/3 \rfloor} < 1.59^n,$$

and thus

$$\sum_{i=1}^{\lfloor n/3 \rfloor} f(i) = \mathcal{O}(1.59^n). \quad (5.7)$$

For $\alpha > 1/3$, we use the fact (Lemma 3.24) that

$$\sum_{k=0}^{\lfloor j/2 \rfloor} \binom{j-k}{k} = F(j+1),$$

where

$$F(j+1) = \left\lfloor \frac{\varphi^{j+1}}{\sqrt{5}} + \frac{1}{2} \right\rfloor$$

is the $(j+1)$th Fibonacci number and $\varphi = (1+\sqrt{5})/2$ is the Golden Ratio. Then

$$\sum_{i=1}^{\lfloor (1-\alpha)n/2 \rfloor} \binom{\lfloor (1+\alpha)n/2 \rfloor}{i} \leq \sum_{i=1}^{\lfloor (1-\alpha)n/2 \rfloor} \binom{n-i}{i}$$

$$\leq \sum_{i=1}^{\lfloor n/2 \rfloor} \binom{n-i}{i} < \varphi^{n+1} < n \cdot 1.6181^n.$$

Therefore,

$$\sum_{i=\lfloor n/3 \rfloor}^{n} f(i) = \mathcal{O}(1.6181^n). \quad (5.8)$$

Finally, the lemma follows from the formulas (5.5), (5.7) and (5.8).   □

Let us remark that by making use of Lemma 5.29, the proof of the previous lemma can be turned into an algorithm enumerating all minimal separators within time $\mathcal{O}(1.6181^n)$. Another approach to enumerate all minimal separators is to use an algorithm of Berry, Bordat, and Cogis [19] listing all minimal separators of an input graph G in $\mathcal{O}(n^3 |\Delta_G|)$ time, which by Lemma 5.30, is $\mathcal{O}(1.6181^n)$.

While the enumeration of potential maximal cliques is again based on Lemma 5.28, we need deeper insight in the combinatorial structure of potential maximal cliques.

We need the following combinatorial result of Fomin and Villanger [96].

Lemma 5.31. *For every potential maximal clique Ω of $G = (V, E)$, there exists a vertex set $Z \subseteq V$ and $z \in Z$ such that*

- $|Z| - 1 \le (2/3)(n - |\Omega|)$,
- $G[Z]$ *is connected, and*
- $\Omega = N(Z \setminus \{z\})$ *or* $\Omega = N(Z) \cup \{z\}$.

Now we are in a position to give a bound on the number of potential maximal cliques in a graph.

Lemma 5.32. *Let Π_G be the set of all potential maximal cliques in a graph G on n vertices. Then $|\Pi_G| = \mathcal{O}(1.7549^n)$.*

Proof. By Lemma 5.31, the number of potential maximal cliques of size αn does not exceed the number of connected subsets of size at most $(1 - \alpha)(2n/3)$ times some polynomial of n. Thus the number of potential maximal cliques is

$$\mathcal{O}^* \left(\sum_{i=1}^{\lceil (1-\alpha)(2n/3) \rceil} \binom{\lceil (2+\alpha)n/3 \rceil}{i} \right). \tag{5.9}$$

For $\alpha \le 2/5$, the sum above is upper bounded by $\mathcal{O}^*(2^{\frac{(2+\alpha)n}{3}})$ and for $\alpha \ge 2/5$ by

$$\mathcal{O}^* \left(\binom{\lceil (2+\alpha)n/3 \rceil}{\lceil (1-\alpha)(2n/3) \rceil} \right).$$

By making use of Lemma 3.13, it is possible to show that in both cases the sum in (5.9) is bounded by $\mathcal{O}^*(1.7549^n)$. $\qquad\square$

Bouchitté and Todinca [34, 35] have shown that for a given vertex subset Ω, it is possible to decide in polynomial time whether Ω is a potential maximal clique. Thus by making use of Lemma 5.29, the proof of Lemma 5.32 combined with the recognition algorithm of Bouchitté and Todinca can be turned into an algorithm enumerating all potential maximal cliques in time $\mathcal{O}(1.7549^n)$. These enumeration algorithms, combined with the bounds from Lemmata 5.24 and 5.32 complete the proof of Theorem 5.24

Notes

The treewidth of a graph is a fundamental graph parameter. It was introduced by Halin in [109] and rediscovered by Robertson and Seymour [181, 182, 183] and, independently, by Arnborg and Proskurowski [7]. Many hard optimization problems on graphs are solvable in polynomial time when the treewidth of the input graph is small [54, 53].

The result of Exercise 5.8 was rediscovered several times in terms of search number, vertex separations and pathwidth. See, e.g. the papers of Ellis, Sudborough, and

Turner [68], Petrov [174] and Parsons [170]. The proof of Theorem 5.10 is due to Monien and Preis and can be found in [158]. Lower bounds on pathwidth of cubic graphs can be obtained by making use of Algebraic Graph Theory. In particular, Bezrukov, Elsässer, Monien, Preis, and Tillich [20] (by making use of the second smallest eigenvalues of a Ramanujan graph's Laplacian) showed that there are 3-regular graphs with the bisection width at least $0.082n$. (See [20] for more details.) It can be shown that the result of Bezrukov et al. also yields the lower bound $0.082n$ for pathwidth of graphs with maximum degree three. It is an interesting challenge to reduce the gap between $0.082n$ and $0.167n$.

The proof of Theorem 5.14 is given by Fomin, Gaspers, Saurabh and Stepanov [80]. The proof of Lemma 5.17 is from Diaz, Serna and Thilikos [64]. Parts of the results of Sect. 5.2 are taken from Kneis, Mölle, Richter, and Rossmanith [134]. The proof of Lemma 5.16 can be found in the book of Kloks [132].

In the proof of Theorem 5.18, we mention that one can compute $\mathrm{aut}(F)$ in time $\mathcal{O}(2^{n_F} \cdot n_F{}^{t+2} \cdot t)$ and space $\mathcal{O}(\log n_F \cdot n_F{}^{t+1})$. While this is sufficient for our purposes, let us remark that by the classical result of Babai, Kantor and Luks [9], one can compute $\mathrm{aut}(F)$ for a graph F on n_F vertices in time $2^{\mathcal{O}(\sqrt{n_F \log n_F})}$. In fact, Babai et al. solve the harder problem of computing the automorphism group and its generators for a given graph F.

The proof of Theorem 5.24 can be found in [90]. It is based on ideas from [34, 35] and uses dynamic programming over structures related to potential maximal cliques called blocks. Lemma 5.28 and its usage to bound the number of minimal separators and potential maximal cliques is taken from [95]. A slightly better bound on the number of potential maximal cliques is obtained in [96]. Polynomial space algorithms computing treewidth are given in [31, 95]. An algorithm enumerating minimal separators can be found in [19]. The example in Fig. 5.5 can be generalized to a graph on n vertices (vertices u and v connected by $(n-2)/3$ paths of length 4) such that this graph contains $3^{\frac{n-2}{3}}$ minimal u,v-separators. It is an open question, whether the number of minimal separators in every n-vertex graph is $\mathcal{O}^*(3^{n/3})$.

Chapter 6
Measure & Conquer

Measure & Conquer is a powerful method used to analyse the running time of branching algorithms. Typically it allows us to achieve running times for branching algorithms that seem hard or even impossible to establish by the simple analysis of branching algorithms studied in Chap. 2. The main difference is that the measure for the size of an instance of a subproblem and thus also the measure for the progress during the branching algorithm's execution will be chosen with much more freedom. Conceptually the approach is not very different from simple analysis. Nevertheless the innocent looking change has important consequences. On the technical side simple branching algorithms with few branching and reduction rules become competitive while the analysis of the running time is typically quite complex and computing power is needed to determine a clever choice of the measure. Often coefficients involved in the measure, typically called weights, need to be fixed so as to minimize the established running time.

Most of the currently best known exact algorithms to solve particular NP-hard problems, which are branching algorithms, have been obtained during the last decade using Measure & Conquer or related methods.

The idea behind Measure & Conquer is to focus on the choice of the measure, instead of creating algorithms with more and more branching and reduction rules. Typically a measure should satisfy the following three conditions:

- The measure of an instance of a subproblem obtained by a reduction rule or a branching rule is smaller than the measure of the instance of the original problem.
- The measure of each instance is nonnegative.
- The measure of the input is upper bounded by some function of "natural parameters" of the input.

The last property is needed to retranslate the asymptotic upper bound in terms of the measure into an upper bound in terms of some natural parameters for the size of the input (such as the number of vertices in a graph or the number of variables in a formula). In this way one is able to derive from different (and often complicated) measures, results that are easy to state and compare.

F.V. Fomin, D. Kratsch, *Exact Exponential Algorithms*, Texts in Theoretical Computer Science. An EATCS Series, DOI 10.1007/978-3-642-16533-7_6,
© Springer-Verlag Berlin Heidelberg 2010

The definitions and results of Section 2.1 will be used throughout this chapter. Real numbers, for example as branching factors, will be represented by rational numbers as described in Section 2.1. The measure of an instance may now be rational, and thus we have rational branching vectors and rational exponents. All this can be treated similarly to Chap. 2, including the computation of branching factors.

6.1 Independent Set

This section provides an introduction to Measure & Conquer. Its goal is to play with measures and to get acquainted with the typical reasonings. Obtaining competitive or best known running times is not the goal. This will be considered in subsequent sections.

Measure & Conquer might look a bit surprising and even counterintuitive at first glance. Actually it is a natural question why a change of the measure should change the established running time. Our goal is twofold. On one hand, we want to illustrate the limits of simple measures, such as those presented in Chap. 2. On the other hand, we want to show why and how more sophisticated measures may provide better upper bounds for the running time of a branching algorithm. We also intend to give some indication of which way to choose measures. Clever measures may exploit particular properties of a branching algorithm that simple measures simply cannot take into account.

To this end we present and study a simple branching algorithm for the MAXIMUM INDEPENDENT SET problem (MIS). The branching algorithm is called mis3 and it is described in Fig. 6.1.

Algorithm mis3(**G**).
Input: A graph $G = (V, E)$.
Output: The maximum cardinality of an independent set of G.

1 **if** $\exists v \in V$ with $d(v) = 0$ **then**
    └ **return** $1 + \text{mis3}(G \setminus v)$

2 **if** $\exists v \in V$ with $d(v) = 1$ **then**
    └ **return** $1 + \text{mis3}(G \setminus N[v])$

3 **if** $\Delta(G) \geq 3$ **then**
    │ choose a vertex v of maximum degree in G
    └ **return** $\max(1 + \text{mis3}(G \setminus N[v]), \text{mis3}(G \setminus v))$

4 **if** $\Delta(G) \leq 2$ **then**
    │ compute $\alpha(G)$ using a polynomial time algorithm
    └ **return** $\alpha(G)$

Fig. 6.1 Algorithm mis3 for MIS

Assume we want to analyze a branching algorithm with the goal to improve upon a trivial enumeration algorithm of running time $\mathcal{O}^*(2^n)$ and to establish a running time $\mathcal{O}^*(c^n)$ with $c < 2$. Clearly it is natural to use as exponent of the running time the measure n of the input size. Therefore simple analysis has been considered as the natural (and only) way to analyze branching algorithms, n being the number of vertices of a graph, the number of variables of a Boolean formula, etc.

Let us be more specific and consider a simple analysis of the algorithm mis3 for the problem MIS. This algorithm consists of three reduction rules. First observe that rule (4) can be performed in polynomial time. Indeed, if every vertex in a graph is of degree at most two, then the graph is the disjoint union of trees and cycles and its treewidth is at most two. A maximum independent set in such graph can be easily found in polynomial time by the results from Chap. 5. Of course, it is possible to solve MIS in polynomial time on trees and cycles without using the treewidth machinery as well.

Algorithm also has one branching rule with branching vector $(1, d(v) + 1)$. Since the algorithm branches only on vertices of degree at least 3, the branching vector has its smallest branching factor if $d(v) = 3$. Clearly for any node of the search tree the running time of all reduction rules executed before the next branching is polynomial, and this is independent of the measure for mis3. Applying reduction rule (1) or (2) removes at least one vertex. Rule (4) will only be applied to an instance in a leaf of the search tree. Hence the running time of mis3 is $\mathcal{O}^*(\alpha^n)$, where α is the unique positive real root of $x^4 - x^3 - 1 = 0$, and thus $\alpha < 1.3803$. Hence $\mathcal{O}(1.3803^n)$ is an upper bound for the worst-case running time of algorithm mis3.

Is it the worst-case running time of mis3? Is it the best one can prove? It is necessary to have a closer look at the analysis and to better understand the algorithm. First let us emphasize that the simple analysis of mis3 means that one assigns a weight of 1 to each vertex of the instance G' which is an induced subgraph of the input graph G. A closer look at the algorithm reveals that before applying the branching rule (3) any instance of a subproblem has the following properties. Any vertex of degree 0 or 1 has been removed from the instance by a reduction rule, either (1) or (2). Hence such vertices do not exist in an instance when branching occurs. However the above simple analysis does not take this into account. Each such reduction decreases the simple weight by 1 or 2 but it is attributed to the reduction, and thus not taken into account for any branching. A vertex of degree 2 will not be removed immediately. It either remains in the instance during the recursive execution until a leaf of the search tree is reached and reduction rule (4) is applied, or it may be removed from an instance by rule (1) or (2) after one of its neighbors has been removed. Finally a vertex of degree at least 3 is one the algorithm might branch on.

Our intuition is the following. The running time of the algorithm is bounded by a polynomial in n times the number of leaves in the search tree, and those leaves essentially are generated by branching rules only. This means that applications of reduction rules are "cheap" and only applications of branching rules are "costly". Hence vertices of degree greater than 3 are expensive while vertices of degree at most 2 seem to be cheap. This suggests that we assign two different weights to vertices, implying the following measure for any instance G' generated by algorithm mis3 when executed on the input graph G:

$$k_1(G') = n_{\geq 3},$$

where $n_{\geq 3}$ is the number of vertices of G' of degree at least 3. This means that vertices of degree at most 2 have weight 0 and vertices of degree at least 3 have weight 1.

Now analyzing the branching rule will become more complicated; in fact more complex measures may lead to an enormous increase in the number of branching vectors and recurrences. The simple analysis of mis3 establishes one branching vector $(1, 1 + d(v))$ for all possible degrees $d(v) \geq 3$. We need to establish the branching vectors with respect to the measure k_1. Say we branch on a vertex v of G' of degree $d \geq 3$. Due to the reduction rules (1) and (2), G' has no vertices of degree 0 or 1. First consider discarding v: the weight of the instance decreases by at least 1, since we remove v. If v has a neighbor w then the degree of w decreases from $d_{G'}(w)$ to $d_{G'-v}(w) = d_{G'}(w) - 1$ and thus the weight of w changes only if $d_{G'}(w) = 3$; it decreases from 1 to 0. Hence when discarding v the weight of the instance decreases by $1 + |\{w : vw \in E, d_{G'}(w) = 3\}|$, which is at least 1. Now consider selecting v: the weight of the instance decreases by the weight of v and all its neighbors, furthermore for every vertex $w \in N_{G'}^2(v)$, i.e. the vertices w at distance 2 from v in G', the degree may decrease and thus their weight may decrease from 1 to 0. This looks for a second like at least the value $d(v) + 1$ of the simple analysis. But no such luck. What if v has neighbors of degree 2? Even worse, what if all neighbors of v have degree 2? Then in the worst case when selecting v there is a decrease of 1 due to the removal of v, and maybe no decrease at all by neighbors or vertices $N_{G'}^2(v)$. We obtain the branching vector $(1, 1)$ and branching factor 2. No doubt our choice of the measure k_1 was not good.

Nevertheless there is an important insight. Vertices of degree 0 and 1 will be removed immediately while vertices of degree 2 remain in the graph, and this is a crucial difference. Thus their weights should be different, and hence we assign weight 0 to vertices of degree at most 1, weight $w_2 \in [0, 1]$ to vertices of degree 2, and weight 1 to vertices of degree at least 3. The new measure for any instance G' generated by algorithm mis3 is

$$k_2(G') = w_2 n_2 + n_{\geq 3},$$

where n_2 is the number of vertices of G' of degree 2 and $n_{\geq 3}$ is the number of vertices of degree at least 3 in G'.

We do not know the best choice of w_2; in general a careful analysis and computing power would be needed to determine the best value of w_2 so as to minimize the corresponding running time. Without such a computational effort we want to argue here that using the new measure one obtains a better bound for the running time. We do not change the algorithm, we simply change the measure, and we establish a better running time. This is the fundamental idea of Measure & Conquer.

Lemma 6.1. *Algorithm* mis3 *has running time* $\mathcal{O}(1.3248^n)$.

Proof. We analyze the running time of algorithm mis3 with respect to the measure k_2 and we fix $w_2 = 0.5$; simply to see what we achieve by this choice. First we determine the branching vectors for branching rule (3). Suppose the algorithm branches on a vertex v of degree $d \geq 3$ in the graph G'. Then $\delta(G') \geq 2$ and $\Delta(G') = d$. Let $u_1, u_2, \ldots u_d$ be the neighbors of v. We denote by OUT the decrease (or a lower bound of the decrease) of the measure of the instance when branching to a subproblem by discarding v. Similarly, we denote by IN the decrease (or a lower bound on the decrease) of the measure of the instance when branching to a subproblem by selecting v. The removal of v decreases the weight by 1 for both subproblems "discard v" and "select v". Furthermore for each vertex u_i of degree 2, its weight decreases by $w_2 = 0.5$ when discarding v, and also when selecting v. Hence the branching vector is (OUT, IN) with $IN \geq 1$, $OUT \geq 1$ and $IN + OUT \geq 2 + d(v)$. Hence $\tau(OUT, IN) \leq \tau(1, 1 + d(v))$ by Lemmata 2.2 and 2.3. Consequently if mis3 branches on a vertex v of degree at least 4 then its branching vector (OUT, IN) has a factor $\tau(OUT, IN) \leq \tau(1, 5) < 1.3248$.

To prove the lemma it remains to reconsider branching on a vertex v with $d_{G'}(v) = 3$. Due to the maximum degree condition in the branching rule, the graph G' has vertices of degree 2 or 3 only. Let w be such a vertex. Discarding w decreases the weight of the instance by 0.5 or 1, and discarding a neighbor of w decreases the weight of w by 0.5. This implies immediately that $OUT \geq 1 + 3 \cdot 0.5$ and $IN \geq 1 + 3 \cdot 0.5$ and $\tau(OUT, IN) \leq \tau(2.5, 2.5) < 1.3196$. Hence all branching factors of the algorithm are upper bounded by 1.3248, which completes the proof of the lemma. □

We have experienced a significant improvement on the established running time by the choice of the measure. Not surprisingly this is not the best choice of the measure. Let us call n_i the number of vertices of degree i, $i \geq 0$, for the instance of a graph problem. Then the following measure will often be a good choice to start with.

$$k(G') = \sum_{i=0}^{n} w_i n_i,$$

where $w_i \in [0, 1]$ for all $i \geq 0$.

When executing algorithm mis3 vertices of degree at most 1 do not exist in an instance mis3 is branching on, and thus $w_0 = w_1 = 0$ is a good choice.

Suppose we consider now a modification of the above measure by setting $w_0 = w_1 = 0$ $0 \leq w_2 \leq w_3 \leq 1$ and $w_t = 1$ for all $t \geq 4$. Then with best choice of the weights $w_2 = 0.596601$, $w_3 = 0.928643$, we obtain a running time of $\mathcal{O}(1.2905^n)$ for algorithm mis3.

Lemma 6.2. *Algorithm* mis3 *has running time* $\mathcal{O}(1.2905^n)$.

Exercise 6.3. Try to improve upon the upper bound of the running time $\mathcal{O}(1.2905^n)$ of algorithm mis3 using Measure & Conquer and a more clever measure, for example more different weights in the measure.

6.2 Feedback Vertex Set

In this section we describe a branching algorithm solving the FEEDBACK VERTEX SET problem in time $\mathcal{O}(1.899^n)$. An interesting feature of this algorithm is that the analysis of its running time crucially depends on the choice of the measure of the instance of a subproblem. Remarkably simple analysis, i.e. measuring the progress of the algorithm in terms of vertices of the graph, provides us with the running time $\mathcal{O}^*(2^n)$, which is the same as by the brute-force algorithm that tries all vertex subsets and checks whether the subset induces a forest.

Feedback Vertex Set Problem. In the FEEDBACK VERTEX SET problem (FVS) we are given an undirected graph $G = (V, E)$. The task is to construct a feedback vertex set $W \subseteq V$ of minimum size. A set of vertices W of a graph $G = (V, E)$ is called a *feedback vertex set* if the graph $G \setminus W$ does not contain any cycle; in other words, $G \setminus W$ is a forest. To solve FVS we solve the problem of finding a maximum size subset $F \subseteq V$ such that $G[F]$ is a forest. This problem is called the MAXIMUM INDUCED FOREST problem. Clearly W is a (minimum) feedback vertex set of G if and only if $V \setminus W$ is a (maximum) induced forest of G.

Let us start with a small example, which explains the main intuition behind the algorithm. Consider the following branching procedure recursively constructing induced forests in a graph G. Suppose that at some step of the algorithm we have constructed a forest F. Now we want to see how this forest can be extended to a larger forest. We take a vertex $v \in V \setminus F$ adjacent to some vertex t of the forest F, and then perform branching in two subproblems: either v is in the new forest or not. In the first case, every neighbor u of v outside F that also has a neighbor in the connected component of F that contains t, cannot be in the forest—adding v and u to F will create a cycle. Thus we can remove all neighbors of v that have neighbors in the component of F that contains t, and add v to the forest. In the second case we just remove v.

A reasonable measure for such a branching rule seems to be the number of vertices not included in the forest. Then the algorithm grows the forest as follows. We start growing the forest from a vertex, and at every step, we take a vertex adjacent to the forest, and either include it in the forest, or remove it from the graph. Suppose that we want to add a vertex v which is adjacent to at least three vertices outside the forest. If one of these neighbors, say u, is also adjacent to some vertex in the same connected component in the forest as v, then u and v cannot simultaneously be in a solution. Thus when we add v to a forest, we must remove u. This case is good for us. However, it can happen that vertex v has 3 neighbors which are not adjacent to vertices from F. If in this case we branch on two subproblems, and thus either add v to the forest, which reduces the number of vertices in the new problem by 1, or remove v from the graph, which again reduces the number of vertices only by 1. The worst-case running time $T(n)$ of the algorithm, up to a polynomial factor, satisfies the following recurrence

$$T(n) \leq T(n-1) + T(n-1).$$

The branching vector is $(1,1)$ and its branching factor $\tau(1,1) = 2$, which implies a running time of $t(n) = \mathcal{O}^*(2^n)$. This is exactly what can be obtained by the brute-force approach. At first glance, this branching algorithm does not give us any advantage.

To obtain a better choice of the measure, the trick is to put different weights on different types of vertices. Let us do a simple thing—assign weight 0 to all vertices that are in the forest, assign weight 1 to all vertices adjacent to vertices of the forest but not in the forest, and weight 1.5 to all remaining vertices. Define the measure μ_G of graph G as the sum of all its vertex weights. Then the measure of G denoted by μ_G satisfies $\mu_G \leq 1.5n$. If we manage to prove that the running time of the algorithm is $\mathcal{O}^*(c^{\mu_G})$ for some constant c, then we can also upper bound the running time of the algorithm by $\mathcal{O}^*(c^{1.5n})$.

Now in our example, assume for a moment that all three neighbors of v are not adjacent to the vertices of the forest. Then when we add v to the forest, we reduce the measure of the new graph by 1 because the weight of v changes from 1 to 0 and by $3/2$ because the weights of the three neighbors of v have been changed from 1.5 to 1. Thus in the subproblem where v is selected in the forest, the total measure of the instance is decreased by 2.5. If we discard v and exclude it from the forest, we simply remove v from the graph. Then the measure of the instance is decreased by 1.

The corresponding recurrence is

$$T(\mu_G) \leq T(\mu_G - 2.5) + T(\mu_G - 1).$$

The branching vector of the recurrence is $(2.5, 1)$ and its branching factor is $\alpha < 1.5290$. Thus the solution of the recurrence is

$$T(\mu_G) = \mathcal{O}(1.5290^{\mu_G}) = \mathcal{O}(1.5290^{1.5n}) = \mathcal{O}(1.8907^n).$$

Of course, to estimate the running time of the algorithm we have to take care of all different cases, like what happens if v has 2 neighbors outside the forest, or if some neighbors of v also have neighbors in the forest. Why does playing with a measure bring us to a better running time; what is the magic here? The answer is that a better choice of measure can improve the recurrence corresponding to the worst case (three neighbors of v do not have neighbors in the forest in our example). Such improvement is not for free—the recurrences corresponding to better cases can become worse, but often by somewhat *balancing the recurrences* we can gain in general.

The moral of this example is that estimating the worst-case running time of a branching algorithm can depend on the way we measure the size of the instances of the problem.

In the remainder of this section we present in detail the algorithm for FVS and its analysis. The main idea of the algorithm follows the example above but several additional issues have to be taken care of.

- The choice of weights $\{0,1,1.5\}$ is quite arbitrarily. (The same running time would be obtained when normalizing the weights, for example by dividing all weights by 1.5.) Can we improve the analysis by choosing other weights? Indeed, we can.
- The cases in which the vertex v has one or two neighbors outside of F need to be handled separately. It turns out that it is possible to avoid branching in those cases.

6.2.1 An Algorithm for Feedback Vertex Set

We call a subset $F \subseteq V$ *acyclic* if $G[F]$ is a forest. If F is acyclic then every connected component of $G[F]$ on at least two vertices is called *non-trivial*. If T is a non-trivial connected component of $G[F]$ then we denote by $\mathtt{compress}(T \to t)$ the compression of T into t which is the following operation

- We contract all edges of T into one vertex t and remove emerging loops. Note that this operation may create multiedges in G.
- We remove all vertices that are adjacent to t by a multiedge.

For example, the graph G in Fig. 6.2 consists of a component T induced by the vertices u, x, y, z and vertex a. After contracting the edges of T, one obtains the graph with the two vertices t and a and an edge of multiplicity two. Thus the result of $\mathtt{compress}(T \to t)$ on G is the singleton graph consisting of vertex t.

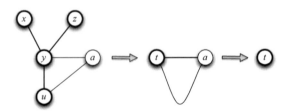

Fig. 6.2 Example of operation $\mathtt{compress}(T \to t)$

For an acyclic subset $F \subseteq V$, denote by $\mathcal{M}_G(F)$ the set of all maximum acyclic supersets of F in G (we omit the subscript G when it is clear from the context which graph is meant). Let $\mathcal{M} := \mathcal{M}(\emptyset)$. Then the problem of finding a maximum induced forest can be stated as finding an element of $\mathcal{M}$. We solve a more general problem, namely finding an element of $\mathcal{M}(F)$ for an arbitrary acyclic subset F.

To simplify the description of the algorithm, we suppose that F is always an independent set. The next lemma justifies this assumption.

Lemma 6.4. *Let $G = (V, E)$ be a graph, $F \subseteq V$ be an acyclic subset of vertices and T be a non-trivial connected component of $G[F]$. Denote by G' the graph obtained from G by* compress$(T \to t)$. *Then*

- $X \in \mathcal{M}_G(F)$ *if and only if* $X \setminus T \cup \{t\} \in \mathcal{M}_{G'}(F \setminus T \cup \{t\})$.

Proof. Let $X \in \mathcal{M}_G(F)$. If after contracting all edges of T to t, a vertex v is adjacent to t by a multiedge, then the set $T \cup \{v\}$ is not acyclic in G. Hence, no element of $\mathcal{M}_G(F)$ contains v. In other words, no vertex of $X \setminus T$ is removed by the compression compress$(T \to t)$. Thus $X' = X \setminus T \cup \{t\}$ is a set of vertices of G'.

We claim that X' is an acyclic subset of G'. For a contradiction, let us assume that X' induces a cycle C' in G'. Then X' contains t because otherwise cycle C' is also a subgraph of $G[X]$, the subgraph of G induced by the acyclic set X, which is a contradiction. Let x_1 and x_2 be the two neighbors of t in C'. There is a path in G from x_1 to x_2 in G such that all inner vertices of this path are in T. Replace t in C' by such a path. As a result we obtain a cycle C induced by X in G, which is a contradiction to the acyclicity of X. Similar arguments show that if X' is an acyclic subset in G' then $X = X' \cup T \setminus \{t\}$ is acyclic in G.

Finally we claim that X' is a maximum acyclic subset of G'. We have that $|X'| = |X| - |T| + 1$. If there is an acyclic subset Y' of G', such that $|Y'| > |X'|$, then the set $Y = Y' \cup T \setminus \{t\}$ is an acyclic subset of G and $|Y| > |X|$. However this contradicts the choice of X. $\qquad\square$

By Lemma 6.4, we can compress every non-trivial component of F. In this way we obtain an equivalent instance where the new vertex set to be extended into a maximum induced forest is an independent set.

The following lemma is used to justify the main branching rule of the algorithm.

Lemma 6.5. *Let $G = (V, E)$ be a graph, $F \subseteq V$ an independent set of G and $v \notin F$ a vertex adjacent to exactly one vertex $t \in F$. Then there exists $X \in \mathcal{M}(F)$ such that either v or at least two vertices of $N(v) \setminus \{t\}$ are in X.*

Proof. If for some $X \in \mathcal{M}(F)$, $v \notin X$, and no vertex of $N(v) \setminus \{t\}$ is in X, then $X \cup \{v\}$ is also an induced forest of G of size larger than $|X|$. Thus if $v \notin X$, then at least one vertex $z \in N(v) \setminus \{t\}$ is in X. If z is not the only such vertex, then the lemma follows.

Let us assume that z is the only vertex in $N(v) \setminus \{t\}$ from X. Since X is maximal, we have that $X \cup \{v\}$ is not acyclic. Because v is of degree at most 2 in $G[X \cup \{v\}]$, we conclude that all cycles in $G[X \cup \{v\}]$ must contain z. Then the set $X \cup \{v\} \setminus \{z\}$ is acyclic, of maximum size and satisfies the condition of the lemma. $\qquad\square$

6.2.2 Computing a Minimum Feedback Vertex Set

Now everything is ready to give the description of the algorithm. Instead of computing a minimum feedback vertex set directly, the algorithm finds the maximum size of an induced forest in a graph. In fact, it solves a more general problem: for any acyclic set F it finds the maximum size of an induced forest containing F. Let us remark, that the algorithm can easily be turned into an algorithm computing such a set (instead of its cardinality only).

During an execution of the algorithm one vertex $t \in F$ is called the *active vertex*. The algorithm branches on a chosen neighbor of the active vertex t. Let $v \in N(t)$. Denote by K the set of all vertices of F other than t that are adjacent to v. Let G' be the graph obtained after the compression $\mathtt{compress}(K \cup \{v\} \rightarrow u)$. We say that a vertex $w \in V \setminus \{t\}$ is a *generalized neighbor* of v in G if w is a neighbor of u in G'. Denote by $\mathrm{gd}(v)$ the *generalized degree* of v which is the number of its generalized neighbors.

For example, in Fig. 6.3, say we have $K = \{t'\}$. After the compression $\mathtt{compress}$ $(K \cup \{v\} \rightarrow u)$ in the new graph the neighbors of u (except t) are y, z, and w. Thus the generalized degree of v is 3.

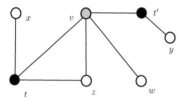 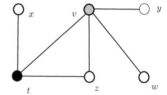

Fig. 6.3 Example of compression and generalised degree

The description of the algorithm consists of a sequence of cases and subcases. To avoid a confusing nesting of if-then-else statements let us use the following convention: the first case which applies is used in the algorithm. Thus, inside a given case, the hypotheses of all previous cases are assumed to be false.

Algorithm $\mathtt{mif}(G,F)$ computes for a given graph $G = (V,E)$ and an acyclic set $F \subseteq V$ the maximum size of an induced forest containing F. It is described by the following preprocessing and main procedures. Let us note that $\mathtt{mif}(G,\emptyset)$ computes the maximum size of an induced forest in G.

Preprocessing

1. If G consists of $k \geq 2$ connected components $G_1, G_2, \ldots, G_k$, then the algorithm is called on each of the components and

$$\mathtt{mif}(G,F) = \sum_{i=1}^{k} \mathtt{mif}(G_i, F_i),$$

where $F_i := V(G_i) \cap F$ for all $i \in \{1, 2, \ldots, k\}$.
2. If F is not independent, then apply operation $\mathtt{compress}(T \rightarrow v_T)$ on an arbitrary non-trivial component T of F. If T contains the active vertex then v_T becomes active. Let G' be the resulting graph and let F' be the set of vertices of G' obtained from F. Then

$$\mathtt{mif}(G,F) = \mathtt{mif}(G',F') + |T| - 1.$$

Main procedures

1. If $F = V$, then $\mathcal{M}_G(F) = \{V\}$. Thus,

$$\mathtt{mif}(G,F) = |V|.$$

2. If $F = \emptyset$ and $\Delta(G) \leq 1$, then $\mathcal{M}_G(F) = \{V\}$ and

$$\mathtt{mif}(G,F) = |V|.$$

3. If $F = \emptyset$ and $\Delta(G) \geq 2$, then the algorithm chooses a vertex t in G of degree at least 2. The algorithm branches on two subproblems—either t is contained in a maximum induced forest, or not—and returns the maximum:

$$\mathtt{mif}(G,F) = \max\{\mathtt{mif}(G,F \cup \{t\}),$$
$$\mathtt{mif}(G \setminus \{t\},F)\}.$$

4. If F contains no active vertex, then choose an arbitrary vertex $t \in F$ as an active vertex. Denote the active vertex by t from now on.
5. If there is a vertex $v \in N(t)$ with $\mathrm{gd}(v) \leq 1$, then add v to F:

$$\mathtt{mif}(G,F) = \mathtt{mif}(G,F \cup \{v\}).$$

6. If there is a vertex $v \in N(t)$ with $\mathrm{gd}(v) \geq 3$, then branch into two subproblems by either selecting v in F or discarding v from F and removing v from G:

$$\mathtt{mif}(G,F) = \max\{\mathtt{mif}(G,F \cup \{v\}),$$
$$\mathtt{mif}(G \setminus \{v\},F)\}.$$

7. If there is a vertex $v \in N(t)$ with $\mathrm{gd}(v) = 2$ then denote its generalized neighbors by w_1 and w_2. Branch by either selecting v in F, or by discarding v from F and removing it from G and also select w_1 and w_2 in F. If selecting w_1 and w_2 to F creates a cycle, just ignore the second subproblem.

$$\mathtt{mif}(G,F) = \max\{\mathtt{mif}(G,F \cup \{v\}),$$
$$\mathtt{mif}(G \setminus \{v\},F \cup \{w_1,w_2\})\}.$$

Thus the algorithm has three branching rules (cases *Main 3, 6* and *7*). The correctness and the running time of the algorithm are analyzed in the following.

Theorem 6.6. *The problem* FEEDBACK VERTEX SET *is solvable in time* $\mathcal{O}(1.8899^n)$.

Proof. Let G be a graph on n vertices. We consider the algorithm $\mathtt{mif}(G,F)$ described above. In what follows, we prove that a maximum induced forest of G can be computed in time $\mathcal{O}(1.8899^n)$ by algorithm $\mathtt{mif}(G,\emptyset)$. The correctness of *Preprocessing 1* and *Main 1,2,3,4,6* is clear. The correctness of *Preprocessing 2* follows from Lemma 6.4. The correctness of cases *Main 5,7* follows from Lemma 6.5 (indeed, applying Lemma 6.5 to the vertex u of the graph G' shows that for some $X \in \mathcal{M}_G(F)$ either v, or at least two of its generalized neighbors are in X).

In order to evaluate the time complexity of the algorithm we use the following measure of an instance (G, F) of a subproblem:

$$\mu(G, F) = |V \setminus F| + c|V \setminus (F \cup N(t))|$$

where the value of the constant c is to be determined later. In other words, each vertex in F has weight 0, each vertex in $N(t)$ has weight 1, each other vertex has weight $1 + c$, and the size of the instance is equal to the sum of the vertex weights. Let us note that

$$\mu(G, F) \leq (1 + c)n.$$

Thus if we prove that the running time of the algorithm is $\mathcal{O}^*(\beta^\mu)$ for some constant β, then we can estimate the running time by $\mathcal{O}^*(\beta^{(1+c)n})$.

Each of the following (non-branching) reduction steps reduces the size of the problem in polynomial time, and has no influence on the base of the running time of the algorithm: *Preprocessing 1,2* and *Main 1,2,4,5* as discussed in Chap. 1.

In all the remaining cases the algorithm branches into smaller subproblems. We consider these cases separately.

In the case *Main 3* every vertex has weight $1 + c$. Therefore, removing v leads to a problem of size $\mu - 1 - c$. Otherwise, v becomes active after the next Main 4 step. Then all its neighbors become of weight 1, and we obtain a problem of size at most $\mu - 1 - 3c$ because v has degree at least 2. This corresponds to the following recurrence

$$T(\mu) \leq T(\mu - 1 - c) + T(\mu - 1 - 3c).$$

In the case *Main 6* removing the vertex v decreases the size of the problem by 1. If v is added to F then we obtain a non-trivial component in F, which is contracted into a new active vertex t' at the next Preprocessing 2 step. Those of the generalized neighbors of v that are of weight 1 will be connected to t' by multiedges and thus removed during the next Preprocessing 2 step. If a generalized neighbor of v is of weight $1 + c$, then it will become a neighbor of t', i.e. of weight 1. Thus, in any case the size of the problem is decreased by at least $1 + 3c$. So, we have that in this case

$$T(\mu) \leq T(\mu - 1) + T(\mu - 1 - 3c).$$

In the case *Main 7* we distinguish three subcases depending on the weights of the generalized neighbors of v. Let i be the number of generalized neighbors of v of weight $1 + c$. Adding v to F reduces the weight of a generalized neighbor either from 1 to 0 or from $1 + c$ to 1. Removing v from the graph reduces the weight of both generalized neighbors of v to 0 (since we add them to F). According to this, we obtain three recurrences, for each $i \in \{0, 1, 2\}$,

$$T(\mu) \leq T(\mu - (3 - i) - ic) + T(\mu - 3 - ic).$$

In total we have established five recurrences. For each fixed value of c we compute the largest branching factor β_c among all of the five branching factors. Then for a fixed value of c, the running time of the algorithm can be bounded by $\mathcal{O}(\beta_c^\mu \mu^{\mathcal{O}(1)})$. Thus everything boils down to searching for all $0 \leq c \leq 1$ for the maximum branching factor β_c among all five recurrences, and then finding the value of c to minimize β_c. By the use of a computer program one can find this optimal value of c. By putting $c = 0.565$, we find $\beta_c < 1.51089$. Thus the running time of our algorithm is

$$T(n) = \mathcal{O}(1.51089^{(1+c)n}) = \mathcal{O}(1.8899^n).$$

$\square$

6.3 Dominating Set

In this section we show how to use a "Measure & Conquer" based algorithm to compute a minimum dominating set of a graph. An important idea is to reduce the original MINIMUM DOMINATING SET problem (MDS) to an equivalent MINIMUM SET COVER problem (MSC), and to use a simple branching algorithm to solve the MSC problem. To establish the running time of this algorithm an involved Measure & Conquer analysis is used. Remarkably, Measure & Conquer improves the running time established for the algorithm from $\mathcal{O}(1.9052^n)$ to $\mathcal{O}(1.5259^n)$.

Let us recall the following definitions.

Minimum Dominating Set Problem. In the MINIMUM DOMINATING SET problem (MDS) we are given an undirected graph $G = (V, E)$. The task is to compute the minimum cardinality of a dominating set D of G, i.e. D is a vertex subset such that every vertex of G is either in D, or adjacent to some vertex in D.

Minimum Set Cover Problem. In the MINIMUM SET COVER problem (MSC) we are given a universe $\mathcal{U}$ of elements and a collection $\mathcal{S}$ of (non-empty) subsets of $\mathcal{U}$. The task is to compute the minimum cardinality of a set cover of $(\mathcal{U}, \mathcal{S})$. A set cover of $(\mathcal{U}, \mathcal{S})$ is a subset $\mathcal{S}' \subseteq \mathcal{S}$ which covers $\mathcal{U}$, i.e. $\bigcup_{S \in \mathcal{S}'} S = \mathcal{U}$. Sometimes we also speak of a set cover of $\mathcal{S}$, instead of a set cover of $(\mathcal{U}, \mathcal{S})$.

The minimum dominating set algorithm is based on a natural reduction of the MINIMUM DOMINATING SET problem to an equivalent MINIMUM SET COVER problem, which has already been used in Sect. 3.2: An input $G = (V, E)$ of MDS is assigned to an input $(\mathcal{U}, \mathcal{S})$ of MSC by imposing $\mathcal{U} = V$ and $\mathcal{S} = \{N[v] \mid v \in V\}$. Note that $N[v]$ is the set of vertices dominated by v, thus D is a dominating set of G if and only if $\{N[v] \mid v \in D\}$ is a set cover of $\{N[v] \mid v \in V\}$. Thus every minimum set cover of $\{N[v] \mid v \in V\}$ corresponds to a minimum dominating set of G. It is crucial that in any input $(\mathcal{U}, \mathcal{S})$ of MSC corresponding to an input $G = (V, E)$ of MDS, it holds that $|\mathcal{U}| = |\mathcal{S}| = |V| = n$.

6.3.1 The Algorithm msc

We consider a simple branching algorithm msc for solving the MSC problem, described in Fig. 6.4. Its output $\text{msc}(\mathcal{S})$ is the minimum cardinality of a set cover of the input $\mathcal{S}$.

A central notion is the *frequency* of an element $u \in \mathcal{U}$ defined to be the number of subsets $S \in \mathcal{S}$ containing u. For the sake of simplicity, we assume that $\mathcal{S}$ covers $\mathcal{U}$: $\mathcal{U} = \mathcal{U}(\mathcal{S}) \triangleq \cup_{S \in \mathcal{S}} S$. With this assumption, an instance of msc is univocally specified by $\mathcal{S}$.

The following lemma justifies some of the reduction rules of algorithm msc.

Lemma 6.7. *For a given MSC instance $\mathcal{S}$:*

1. *If there are two distinct sets S and R in $\mathcal{S}$, $S \subseteq R$, then there is a minimum set cover which does not contain S.*
2. *If there is an element u of $\mathcal{U}$ which belongs to a unique $S \in \mathcal{S}$, then S belongs to every set cover.*

Exercise 6.8. We leave the proof of Lemma 6.7 as an exercise to the reader.

Each set $S \in \mathcal{S}$ of cardinality one satisfies exactly one of the properties in Lemma 6.7, and thus it can be removed by a reduction rule.

The next lemma justifies the final reduction rule of algorithm msc. We need the following definitions. A set $A \subseteq E$ of edges of a graph $G = (V, E)$ is an *edge cover*, if every vertex of G is endpoint of an edge of A; the edge set A is a *matching* if no vertex of G is an endpoint of two edges of A.

Lemma 6.9. *For a given MSC instance $\mathcal{S}$ such that all the subsets S of $\mathcal{S}$ are of cardinality two, MSC can be solved in polynomial time.*

Proof. If all the subsets of $\mathcal{S}$ are of cardinality two then MSC can be solved in polynomial time via the following standard reduction to the MAXIMUM MATCHING problem. Consider the graph $\tilde{G}$ which has a vertex u for each $u \in \mathcal{U}$, and an edge $\{u, v\}$ for each subset $S = \{u, v\}$ in $\mathcal{S}$. Thus we have to compute a minimum edge cover of $\tilde{G}$. To compute a minimum edge cover of $\tilde{G}$, first we compute a maximum matching M on $\tilde{G}$ in polynomial time. Then, for each unmatched vertex u, we add to M an arbitrary edge which has u as an endpoint (if no such edge exists, there is no set cover at all). Finally the subsets corresponding to M form a minimum set cover of $\mathcal{S}$. □

The algorithm msc is given in Fig. 6.4.
indexalgorithm!msc
If $|\mathcal{S}| = 0$ (line 1), $\text{msc}(\mathcal{S}) = 0$. Otherwise (line 2 and line 3) the algorithm tries to reduce the size of the problem without branching, by applying one of the Properties 1 and 2 of Lemma 6.7. Specifically, if there are two sets S and R, $S \subseteq R$, we have

---

Algorithm msc(S).
Input: A collection S of subsets of a universe U.
Output: The minimum cardinality of a set cover of S.

1 **if** $|S| = 0$ **then**
    └ **return** 0
2 **if** $\exists S, R \in S$ with $S \subseteq R$ **then**
    └ **return** msc($S \setminus \{S\}$)
3 **if** $\exists u \in U(S)$ such that there is a unique $S \in S$ with $u \in S$ **then**
    └ **return** $1 + \text{msc}(\text{del}(S, S))$
4 choose a set $S \in S$ of maximum cardinality
5 **if** $|S| = 2$ **then**
    └ **return** poly-msc(S)
6 **if** $|S| \geq 3$ **then**
    └ **return** $\min(\text{msc}(S \setminus \{S\}), 1 + \text{msc}(\text{del}(S, S)))$

Fig. 6.4 Algorithm msc for MSC

$$\text{msc}(S) = \text{msc}(S \setminus S).$$

If there is an element u which is contained in a unique set S, we have

$$\text{msc}(S) = 1 + \text{msc}(\text{del}(S, S)),$$

where

$$\text{del}(S, S) = \{Z \mid Z = R \setminus S \neq \emptyset, R \in S\}$$

is the instance of msc which is obtained from S by removing the elements of S from the subsets in S, and finally removing the empty sets obtained.

If neither of the two properties above applies, the algorithm takes (line 4) a set $S \in S$ of maximum cardinality. If $|S| = 2$ (line 5), the algorithm directly solves the problem with the polynomial time algorithm poly-msc based on the reduction to maximum matching given in Lemma 6.9. Otherwise (line 6), it branches on the two subproblems $S_{IN} = \text{del}(S, S)$ (the case where S belongs to the minimum set cover) and $S_{OUT} = S \setminus S$ (the case where S is not in the minimum set cover). Thus

$$\text{msc}(S) = \min\{\text{msc}(S \setminus \{S\}), 1 + \text{msc}(\text{del}(S, S))\}.$$

In many branching algorithms any instance of a subproblem either contains a corresponding partial solution explicitly or such a partial solution can easily be attached to the instance. In algorithm msc it is easy to attach the collection of all selected sets of S to a subproblem. Thus the given algorithm msc computing the minimum cardinality of a set cover can easily be modified so that it also provides a minimum set cover.

To illustrate the power of the Measure & Conquer analysis let us first consider the following analysis based on a simple measure $k(S')$ of the size of an instance S' of MSC,

$$k(\mathcal{S}') = |\mathcal{S}'| + |\mathcal{U}(\mathcal{S}')|.$$

Let $\ell(k)$ be the number of leaves in the search tree generated by the algorithm to solve a problem of size $k = k(\mathcal{S})$. If one of the conditions of lines 2 and 3 is satisfied, $\ell(k) \leq \ell(k-1)$. Let S be the set selected in line 4. If $|S| = 2$, then the algorithm directly solves the problem in polynomial time ($\ell(k) = 1$). Otherwise ($|S| \geq 3$), the algorithm branches into the two subproblems $\mathcal{S}_{OUT} = \mathcal{S} \backslash \{S\}$ and $\mathcal{S}_{IN} = \text{del}(\mathcal{S}, S)$. The size of $\mathcal{S}_{OUT}$ is $k-1$ (one set removed from $\mathcal{S}$). The size of $\mathcal{S}_{IN}$ is at most $k-4$ (one set removed from $\mathcal{S}$ and at least three elements removed from $\mathcal{U}$). This brings us to $\ell(k) \leq \ell(k-1) + \ell(k-4)$. We conclude that $\ell(k) \leq \alpha^k$, where $\alpha < 1.3803$ is the (unique) positive root of the polynomial $x^4 - x^3 - 1$. It turns out that the total number of subproblems solved is within a polynomial factor of $\ell(k)$. Moreover, solving each subproblem takes polynomial time. Thus the time complexity of algorithm msc is $\mathcal{O}^*(\ell(k)) = \mathcal{O}^*(\alpha^k) = \mathcal{O}(1.3803^{|\mathcal{S}|+|\mathcal{U}|})$. Thus the corresponding algorithm solving MDS has running time $\mathcal{O}^*((\alpha^2)^n) = \mathcal{O}(1.9052^n)$.

Using Measure & Conquer we will show that the running time of the very same MDS branching algorithm is indeed $\mathcal{O}(1.5259^n)$. This illustrates the power of Measure & Conquer and it also shows that it is worth the effort to study and apply this method.

6.3.2 A Measure & Conquer Analysis

In this section we show how to apply Measure & Conquer to refine the running time analysis and to establish a running time of $\mathcal{O}(1.2353^{|\mathcal{S}|+|\mathcal{U}|})$. This includes a more careful set up of the measure of an instance $\mathcal{S}$ of MSC (including various weights to be fixed later), the analysis of the reduction and branching rules of the algorithm with respect to this measure, and the computation to optimize the weights and the measure.

The type of measure we are going to describe is often useful when analysing branching algorithms to solve NP-hard problems exactly. The original motivation for this choice of the measure is the following. Removing a large set has a different impact on the "progress" of the algorithm msc than removing a small one. In fact, when we remove a large set, we decrease the frequency of many elements. Decreasing elements' frequencies pays of in the long term, since the elements of frequency one can be filtered out (without branching). A dual argument holds for the elements. Removing an element of high frequency is somehow preferable to removing an element of small frequency. In fact, when we remove an element occurring in many sets, we decrease the cardinality of all such sets by one. This is good in the long term, since sets of cardinality one can be filtered out. Both phenomena are not taken into account by the simple measure discussed in the previous subsection.

This suggests the idea to give a different "weight" to sets of different cardinality and to elements of different frequency. In particular, let n_i denote the number of subsets $S \in \mathcal{S}$ of cardinality i. Moreover let m_j denote the number of elements $u \in \mathcal{U}$

of frequency j. The following is the measure $k = k(\mathcal{S})$ of the size of an instance $\mathcal{S}$ of MSC:

$$k(\mathcal{S}) = \sum_{i \geq 1} w_i n_i + \sum_{j \geq 1} v_j m_j,$$

where the weights $w_i, v_j \in [0, 1]$ will be fixed later. Note that $k(\mathcal{S}) \leq |\mathcal{S}| + |\mathcal{U}|$. This guarantees that a running time bound with respect to the measure $k(\mathcal{S})$ directly translates into one with respect to $|\mathcal{S}| + |\mathcal{U}|$ and to $2n$, since for all $\alpha \geq 1$, $\alpha^{k(\mathcal{S})} \leq \alpha^{|\mathcal{S}|+|\mathcal{U}|} \leq \alpha^{2n}$.

Theorem 6.10. *Algorithm* msc *solves the* MINIMUM SET COVER *problem in time* $\mathcal{O}(1.2353^{|\mathcal{U}|+|\mathcal{S}|})$.

Proof. As usual in branching algorithms the correctness of the algorithm follows from the correctness of all reduction and branching rules and is easy to obtain. The following assumptions simplify the running time analysis and they are an important part of the definition of the measure.

$(a) w_i \leq w_{i+1}$ and $v_i \leq v_{i+1}$ for $i \geq 1$;
$(b) w_1 = v_1 = 0$;
$(c) w_i = v_i = 1$ for $i \geq 6$.

The first assumption says that the weights are non-decreasing with the cardinality (frequency). Intuitively, this makes sense: the smaller the cardinality (frequency) the closer is the subset (element) to the state when it will be filtered out (without branching). The second assumption is due to the fact that sets of cardinality one and elements of frequency one can be removed without branching. The last assumption is simply due to the fact that in this analysis weights are chosen to be of value at most 1. It also avoids dealing with an unbounded number of weights in the computations to be described later.

The following quantities turn out to be useful in the analysis:

$$\Delta w_i = w_i - w_{i-1}, \; i \geq 2 \quad \text{and} \quad \Delta v_i = v_i - v_{i-1}, \; i \geq 2,$$

Intuitively, Δw_i (Δv_i) is the reduction of the size of the problem corresponding to the reduction of the cardinality of a set (of the frequency of an element) from i to $i - 1$. We make one last assumption

$(d) \Delta w_i \geq \Delta w_{i+1}$, for $i \geq 2$,

that is the w_i's are increasing at decreasing speed. On the one hand, experiments have shown that this is a reasonable choice, on the other hand the assumption simplifies the subsequent analysis significantly.

Having chosen the measure with 8 weights still to be fixed, we now analyse all reduction and branching rules of the algorithm msc with respect to this measure, similar to the analysis of branching algorithms in Chap. 2.

Let $\ell(k)$ be the number of leaves in the search tree generated by the algorithm to solve a problem of measure k. Clearly, $\ell(0) = 1$. Consider the case $k > 0$ (which implies $S \neq \emptyset$). If one of the conditions of lines 3 and 4 holds, one set S is removed from $\mathcal{S}$. Thus we get $\ell(k) \leq \ell(k - w_{|S|})$, where $w_{|S|} \geq 0$ by assumptions (a) and (c).

Otherwise, let S be the subset selected in line 5. If $|S| = 2$, no subproblem is generated ($\ell(k) = 1$). Otherwise ($|S| \geq 3$), msc generates two subproblems $\mathcal{S}_{IN} = \text{del}(S, \mathcal{S})$ and $\mathcal{S}_{OUT} = \mathcal{S} \backslash S$.

Consider the subproblem $\mathcal{S}_{OUT}$. The size of $\mathcal{S}_{OUT}$ decreases by $w_{|S|}$ because of the removal of S. Let r_i be the number of elements of S of frequency i. Note that there cannot be elements of frequency 1. Hence

$$\sum_{i \geq 1} r_i = \sum_{i \geq 2} r_i = |S|.$$

Consider an element $u \in S$ of frequency $i \geq 2$. When we remove S, the frequency of u decreases by one. As a consequence, the size of the subproblem decreases by Δv_i. Thus the overall reduction of the size of $\mathcal{S}_{OUT}$ due to the reduction of the frequencies is at least

$$\sum_{i \geq 2} r_i \Delta v_i = \sum_{i=2}^{6} r_i \Delta v_i,$$

where we used the fact that $\Delta v_i = 0$ for $i \geq 7$ (assumption (c)).

Suppose that $r_2 > 0$, and let $R_1, R_2, \ldots, R_h$, $1 \leq h \leq r_2$, be the sets distinct from S which share at least one element of frequency 2 with S. When we discard S, we must later select all the sets R_i. Suppose R_i shares $r_{2,i}$ such elements with S. Then $|R_i| \geq r_{2,i} + 1$, since otherwise we would have $R \subseteq S$, which is excluded by line 3. Note that $r_{2,i} < |S|$, since S is of maximum cardinality by algorithm msc. Thus, by assumption (a), the reduction of the size of the problem due to the removal of R_i is $w_{|R_i|} \geq w_{r_{2,i}+1}$. We also observe that, by selecting the R_i's, we remove at least one element $f \notin S$, thus gaining an extra $v_{|f|} \geq v_2$ (here we use assumption (a) again). By a simple case analysis, which we present here in a slightly weakened form, the total reduction of the size of the problem due to the removal of the R_i's is at least

$$\Delta k' = \begin{cases} 0 & \text{if } r_2 = 0; \\ v_2 + w_2 & \text{if } r_2 = 1; \\ v_2 + \min\{2w_2, w_3\} = w_3 & \text{if } r_2 = 2; \\ v_2 + \min\{3w_2, w_2 + w_3\} = w_2 + w_3 & \text{if } r_2 = 3, |S| = 3; \\ v_2 + \min\{3w_2, w_2 + w_3, w_4\} = w_4 & \text{if } r_2 \geq 3, |S| \geq 4. \end{cases}$$

where we used the fact that $\min\{2w_2, w_3\} = w_3$ and $\min\{w_2 + w_3, w_4\} = w_4$ by assumptions (b) and (d).

Consider now the subproblem $\mathcal{S}_{IN}$. The size of $\mathcal{S}_{IN}$ decreases by $w_{|S|}$ because of the removal of S. Let $r_{\geq i} = \sum_{j \geq i} r_j$ be the number of elements of S of frequency at least i. Consider an element $u \in S$ of frequency i ($i \geq 2$). The size of $\mathcal{S}_{IN}$ further decreases by v_i because of the removal of u. Thus the overall reduction due to the removal of the elements u of S is

$$\sum_{i\geq 2} r_i v_i = \sum_{i=2}^{6} r_i v_i + r_{\geq 7},$$

where we used the fact that $v_i = 1$ for $i \geq 7$ (assumption (c)). Let R be a set sharing an element u with S. Note that $|R| \leq |S|$. By removing u, the cardinality of R is reduced by one. This implies a reduction of the size of $\mathcal{S}_{IN}$ by $\Delta w_{|R|} \geq \Delta w_{|S|}$ (assumption (d)). Thus the overall reduction of $\mathcal{S}_{IN}$ due to the reduction of the cardinalities of the sets R is at least:

$$\Delta w_{|S|} \sum_{i\geq 2}(i-1)r_i \geq \Delta w_{|S|} \left(\sum_{i=2}^{6}(i-1)r_i + 6 \cdot r_{\geq 7} \right).$$

Note that this quantity is 0 for $|S| \geq 7$. Putting it all together, for all the possible values of $|S| \geq 3$ and of the r_i's such that

$$\sum_{i=2}^{6} r_i + r_{\geq 7} = |S|,$$

we have the following set of recurrences

$$\ell(k) \leq \ell(k - \Delta k_{OUT}) + \ell(k - \Delta k_{IN}),$$

where

- $\Delta k_{OUT} \triangleq w_{|S|} + \sum_{i=2}^{6} r_i \Delta v_i + \Delta k'$,
- $\Delta k_{IN} \triangleq w_{|S|} + \sum_{i=2}^{6} r_i v_i + r_{\geq 7} + \Delta w_{|S|} \left(\sum_{i=2}^{6}(i-1)r_i + 6 \cdot r_{\geq 7} \right).$

All that remains is to determine the best choice of the 8-tuple

$$(w_2, w_3, w_4, w_5, v_2, v_3, v_4, v_5)$$

such that all simplifying assumptions are satisfied and the corresponding running time is as small as possible. Since $\Delta w_{|S|} = 0$ for $|S| \geq 7$, we have that each recurrence with $|S| \geq 8$ is *dominated* by some recurrence with $|S| = 7$, i.e. for each recurrence with $|S| \geq 8$ (whatever the choice of the weights) there is a recurrence with $|S| = 7$ having larger or equal branching factor. Hence it is sufficient to restrict ourselves to the recurrences for the cases $3 \leq |S| \leq 7$. Thus we consider a large but finite number of recurrences. For every fixed 8-tuple $(w_2, w_3, w_4, w_5, v_2, v_3, v_4, v_5)$ the quantity $\ell(k)$ is upper bounded by α^k, where α is the largest number from the set of real roots of the set of equations

$$\alpha^k = \alpha^{k - \Delta k_{OUT}} + \alpha^{k - \Delta k_{IN}}$$

corresponding to all different combinations of the values of $|S|$ and $r_2, \ldots, r_{|S|}$. Thus the estimation of $\ell(k)$ boils down to choosing the weights to minimize α. This optimization problem is interesting in its own right and we refer to Eppstein's work [72] on quasi-convex programming for a general treatment of such problems.

To find the (nearly) optimal weights the authors used a computer program, based on randomized local search, which turns out to be very fast and sufficiently accurate in practice even for a large number of weights and recurrences. The outcome of the program was:

$$
w_i = \begin{cases} 0.377443 & \text{if } i = 2, \\ 0.754886 & \text{if } i = 3, \\ 0.909444 & \text{if } i = 4, \\ 0.976388 & \text{if } i = 5, \end{cases} \quad \text{and} \quad v_i = \begin{cases} 0.399418 & \text{if } i = 2, \\ 0.767579 & \text{if } i = 3, \\ 0.929850 & \text{if } i = 4, \\ 0.985614 & \text{if } i = 5, \end{cases}
$$

which yields $\alpha < 1.2353$, and thus $\ell(k) = \mathcal{O}(\alpha^k(\mathcal{S}))$. This completes the choice of the measure.

Finally we observe that the time spent on a node of the search tree (without recursive calls) is bounded by a polynomial $poly(n)$ of n, since the time spent on any reduction rule is bounded by a polynomial, and since each reduction rule removes a subset of $\mathcal{S}$ or stops the calculation. Hence the overall running time of the algorithm is $\mathcal{O}(n\,\ell(k)\,poly(n)) = \mathcal{O}^*(\ell(k)) = \mathcal{O}(1.2353^{|\mathcal{U}|+|\mathcal{S}|})$.   □

As already mentioned, MDS can be reduced to MSC by imposing $\mathcal{U} = V$ and $\mathcal{S} = \{N[v]\mid v \in V\}$. The size of the MSC instance with respect to the measure $k(\mathcal{S})$ is at most $2n$. By simply combining this reduction with algorithm msc one obtains:

Corollary 6.11. *The problem* MINIMUM DOMINATING SET *is solvable in time* $\mathcal{O}(1.2353^{2n}) = \mathcal{O}(1.5259^n)$.

Two remarks concerning the computations needed to determine the best choice of all weights should be made. First, given the measure and all weights it is computationally easy to determine the branching factors of all recurrences and verifiy that they are all bounded by a claimed solution α.

Second, there are various ways to compute the best choice of the weights. From a mathematical point of view it is sufficient to mention the weights as a certificate for the claimed value of α. There is no need to explain how those weights have been established.

How to apply a technique called memorization, which produces algorithms using exponential space, and improve upon the running time of a branching algorithm achieved by Measure & Conquer analysis is discussed in Chap. 10.

6.4 Lower Bounds

Branching algorithms are one of the major tools for the design and analysis of exact exponential time algorithms. Their running time analysis has been improved significantly during the last decade. All known methods to analyze the running time of branching algorithms are based on (linear) recurrences. Despite all efforts, establishing the worst-case running time of a branching algorithm seems very difficult;

only for a few branching algorithms is their worst-case running time known. Nowadays all we can establish are upper bounds of the worst-case running time; usually called the running time of the algorithm.

The stated running time of a branching algorithm might (significantly) overestimate its (unknown) worst-case running time. Comparing two algorithms via upper bounds of their worst-case running time is not satisfactory but it is common practice for branching algorithms. The only way out is better methods to analyze branching algorithms.

Lower bounds for the worst-case running time of branching algorithms cannot change this unsatisfactory state. However a lower bound on the worst-case running time of a particular branching algorithm can give an idea how far the current analysis of this algorithm is from being tight. There are in fact important branching algorithms having a large gap between the best known lower and upper bound. Furthermore the study of lower bounds has an interesting side effect; it leads to new insights on particular branching algorithms. The reason is that running time analysis of branching algorithms essentially transforms the algorithm in a collection of recurrences and then solves it; while to achieve lower bounds one needs to study and understand the algorithm and its executions.

We start with yet another simple branching algorithm solving the MIS problem, described in Fig. 6.5.

Algorithm mis4(G).
Input: A graph $G = (V, E)$.
Output: The maximum cardinality of an independent set of G.

> if $\Delta(G) \geq 3$ then
> > choose a vertex v of degree $d(v) \geq 3$ in G
> > **return** $\max(1 + \text{mis4}(G \setminus N[v]), \text{mis4}(G \setminus v))$
>
> if $\Delta(G) \leq 2$ then
> > compute $\alpha(G)$ using a polynomial time algorithm
> > **return** $\alpha(G)$

Fig. 6.5 Algorithm mis4 for MIS

It is not hard to show by simple analysis that the running time of algorithm mis4 is $\mathcal{O}(1.3803^n)$. The question is whether this upper bound is tight or can be decreased, e.g. by using Measure & Conquer similar to algorithm mis3 in Sect. 6.1.

To establish a lower bound for the algorithm mis4 let us consider the sequence of graphs $G_n = (\{1, 2, \ldots, n\}, E_n)$, $n \geq 1$, where two vertices i and j of G_n are adjacent iff $|i - j| \leq 3$, see Fig. 6.6.

Let us consider an execution of mis4 on a graph G_n for sufficiently large n (say $n \geq 10$). The algorithm may choose any vertex of G_n having degree at least 3. To break the tie we may choose any suitable vertex. Hence we use the following tie break rule: "Branch on the smallest vertex of the instance". Inductively, we prove

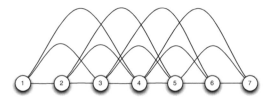

Fig. 6.6 Graph G_7

that every instance generated by mis4 when executed on G_n with the above rule is of the form $G_n[\{i, i+1, \ldots, n\}]$. Hence the smallest vertex is of degree 3 (unless the maximum degree is at most 2). Branching on vertex i of instance $G_n[\{i, i+1, \ldots, n\}]$ calls the instances $G_n[\{i+1, i+2, \ldots, n\}]$, when discarding i, and $G_n[\{i+4, i+5, \ldots, n\}]$, when selecting i.

Let $T(n)$ be the number of leaves in the search tree when running mis4 on G_n with our tie break rule. The above obervations imply the recurrence $T(n) = T(n-1) + T(n-4)$. Let α be the branching factor of the branching vector $(1, 4)$. Then $\Omega(\alpha^n)$ is a lower bound for the worst-case running time of mis4, and this implies

Theorem 6.12. *The worst-case running time of algorithm* mis4 *is* $\Theta^*(\alpha^n)$, *where* $\alpha = 1.3802\ldots < 1.3803$ *is the unique positive real root of* $x^4 - x^3 - 1 = 0$.

Exercise 6.13. Determine a lower bound for the worst-case running time of algorithm mis3 of Sect. 6.1.

The second example is a $\Omega(1.2599^n)$ lower bound on the worst-case running time of algorithm mds of Sect. 6.3. Recall that this algorithm is based on the polynomial-space algorithm msc of Sect. 6.3 and the reduction from MDS to MSC.

Theorem 6.14. *The worst case running time of algorithm* mds *solving the* MINI-MUM DOMINATING SET *problem is* $\Omega(2^{n/3}) = \Omega(1.2599^n)$.

Proof. Consider the following input graph G_n ($n \geq 1$): the vertex set of G_n is $\{a_i, b_i, c_i : 1 \leq i \leq n\}$. The edge set of G_n consists of two types of edges: for each $i = 1, 2 \ldots, n$, the vertices a_i, b_i and c_i induce a triangle T_i; and for each $i = 1, 2, \ldots, n-1$: $\{a_i, a_{i+1}\}$, $\{b_i, b_{i+1}\}$ and $\{c_i, c_{i+1}\}$ are edges, see Fig. 6.7.

Each node of the search tree corresponds to a subproblem of the MSC problem with input $(\mathcal{U}, \mathcal{S} = \{S_v : v \in V\})$ where $S_v \subseteq N[v]$. On the right in Fig. 6.7, we picture the top part of a feasible search tree: there is a node in the tree for each subproblem; subproblems are labelled with the node associated with the branching set; left and right children correspond to selection and discarding of the branching set, respectively.

We give a selection rule for the choice of the vertices v (respectively sets S_v) to be chosen for the branching. Clearly the goal is to choose them such that the number

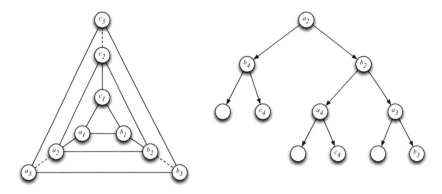

Fig. 6.7 Graph G_n and the feasible search tree

of nodes in the search tree obtained by the execution of algorithm msc on graph G_n is as large as possible.

In each round i, $i \in \{2,3,\ldots,n-1\}$, we start with a pair $P = \{x_i, y_i\}$ of vertices (belonging to triangle T_i), where $\{x_i, y_i\} \subset \{a_i, b_i, c_i\}$. Initially $P = \{a_2, b_2\}$. Our choice makes sure that for each branching vertex x the cardinality of its set S_x is five in the current subproblem $\mathcal{S}$, and that no other rules of the algorithm will apply to a branching vertex than those of line 6 of algorithm msc. Consequently, either the set S_v is taken into the set cover ($\mathcal{S} := \mathrm{del}(\mathcal{S}, S_v)$), or S_v is removed ($\mathcal{S} := \mathcal{S} \setminus S_v$).

For each pair $P = \{x_i, y_i\}$ of vertices we branch in the following 3 ways:
1) take S_{x_i}
2) remove S_{x_i}, and then take S_{y_i}
3) remove S_{x_i}, and then remove S_{y_i}

The following new pairs of vertices correspond to each of the three branches:
1) $P_1 = \{a_{i+2}, b_{i+2}, c_{i+2}\} \setminus x_{i+2}$
2) $P_2 = \{a_{i+2}, b_{i+2}, c_{i+2}\} \setminus y_{i+2}$
3) $P_3 = \{x_{i+1}, y_{i+1}\}$

On each pair P_j we recursively repeat the process. Thus of the three branches of T_i two proceed on T_{i+2} and one proceeds on T_{i+1}.

Let $T(k)$ be the number of leaves in the search tree when all triangles up to T_k have been used for branching. Thus $T(k) = 2 \cdot T(k-2) + T(k-1)$, and hence $T(k) \geq 2^{k-2}$. Consequently the worst case number of leaves in the search tree of msc for a graph on n vertices is at least $2^{n/3-2}$. □

Exercise 6.15. The algorithm of Tarjan and Trojanowski for MIS has been published in 1977 [213] and its (provable) running time is $\mathcal{O}^*(2^{n/3})$. What is the best upper bound of the running time that can be achieved from the linear recurrences provided in [213]? Construct a lower bound for the worst-case running time of this algorithm which should be close to its upper bound.

Exercise 6.16. Provide a lower bound for the worst-case running time of algorithm `mis2` from Chap. 2.

Notes

The presentation of the Measure & Conquer approach is based on the following works of Fomin, Grandoni and Kratsch [83, 85, 87]. This approach was strongly influenced by Eppstein's ideas in quasiconvex analysis [71, 72]. Approaches related to the design and analysis of branching algorithms by Measure & Conquer have been developed by Eppstein (see e.g. [73]), Byskov and Eppstein [43], Scott and Sorkin [203], Gaspers and Sorkin [103] and Dahllöf, Jonsson and Wahlström [58]. Computational issues related to Measure & Conquer are discussed by Williams in [218].

The first algorithm breaking the trivial $\mathcal{O}^*(2^n)$ bound for FEEDBACK VERTEX SET is due to Razgon [177]. Our presentation of the algorithm follows [79] which by a more careful analysis establishes running time $\mathcal{O}(1.7548^n)$. In the same paper it was shown that the number of minimal feedback vertex sets in an n-vertex graph is $\mathcal{O}(1.8638^n)$ and that there are graphs with 1.5926^n different minimal feedback vertex sets. FEEDBACK VERTEX SET is a special case of a more general problem— find a maximum induced subgraph satisfying some property P. Fomin and Villanger [96] used completely different technique based on minimal triangulations of graphs to show that a maximum induced subgraph of treewidth at most t can be found in time $\mathcal{O}(|\Pi_G|n^{\mathcal{O}(t)})$, where $|\Pi_G| = \mathcal{O}^*(1.7348^n)$ is the number of potential maximal cliques in G. For $t \leq 1$, this is gives the fastest so far algorithm for FVS.

Exact algorithms for the NP-complete MINIMUM DOMINATING SET problem can be found in [91, 106, 176, 197]. The algorithm for MINIMUM SET COVER follows [83, 87]. Using the memorization technique (see Chap. 10) the running time of the algorithm solving MINIMUM DOMINATING SET can be improved to $\mathcal{O}(1.5137^n)$, however exponential space is needed. This running time has been improved to $\mathcal{O}(1.5063^n)$ by van Rooij and Bodlaender [187] using Measure & Conquer.

Many branching algorithms solving NP-hard problems exactly have been established in the last five years and a large part of them uses a Measure & Conquer analysis. To mention just a few of them [77, 86, 89, 80, 133, 141, 142, 143, 175, 188]. The PhD theses of Gaspers [101], Stepanov [209] and Liedloff [151] contain more applications of Measure & Conquer.

Chapter 7
Subset Convolution

Subset convolution is a powerful tool for designing exponential time algorithms. The fast subset convolution algorithm computes the convolution of two given functions in time $\mathcal{O}^*(2^n)$, while a direct calculation of such a convolution needs time $\Omega(3^n)$.

In the first two sections we explain the fundamentals of subset convolution and the fast algorithm to compute it. To obtain a fast subset convolution algorithm one relies on repeated use of dynamic programming, and in particular on the so-called fast zeta transform. In the latter sections we present various algorithmic applications of fast subset convolution. In this chapter the algorithms (may) operate with large numbers and thus we use the log-cost RAM model to analyze their running times.

Let us start by providing some fundamental notions for this approach. Let $\mathcal{U}$ be a set of n elements, $n \geq 1$, and let us assume that $\mathcal{U} = \{1, 2, \ldots, n\}$. We denote by $2^{\mathcal{U}}$ the set of all subsets of $\mathcal{U}$. Consider two functions $f, g : 2^{\mathcal{U}} \to \mathbb{Z}$.

Definition 7.1. The *subset convolution*, or for short the *convolution* of f and g, denoted by $f * g$, is a function assigning to any $S \subseteq \mathcal{U}$ an integer.

$$(f * g)(S) = \sum_{T \subseteq S} f(T) \cdot g(S \setminus T) \tag{7.1}$$

The following reformulation of the definition of $(f * g)(S)$ for all $S \subseteq \mathcal{U}$ will also be useful.

$$(f * g)(S) = \sum_{\substack{X, Y \subseteq S \\ X \cup Y = S \\ X \cap Y = \emptyset}} f(X) \cdot g(Y) \tag{7.2}$$

The goal of the next two sections is to present a fast exponential time algorithm to compute the subset convolution of two functions $f, g : 2^{\mathcal{U}} \to \mathbb{Z}$.

F.V. Fomin, D. Kratsch, *Exact Exponential Algorithms*, Texts in Theoretical
Computer Science. An EATCS Series, DOI 10.1007/978-3-642-16533-7_7,
© Springer-Verlag Berlin Heidelberg 2010

7.1 Fast zeta Transform

To establish a fast exponential time algorithm for *subset convolution* we rely on fast
zeta transform and its inverse as well as ranked zeta transform and its inverse.

For a universe $\mathcal{U}$, we consider functions from $2^{\mathcal{U}}$ (the family of all subsets of
$\mathcal{U}$) to $\mathbb{Z}$. For such a function $f : 2^{\mathcal{U}} \to \mathbb{Z}$, the *zeta transform* ζ of f is a function
$f\zeta : 2^{\mathcal{U}} \to \mathbb{Z}$ defined by

$$f\zeta(S) = \sum_{X \subseteq S} f(X).$$

One of the interesting features of the zeta transform of f is that, given $f\zeta$, the
function f can be recovered by a formula called *Möbius inversion*.

Lemma 7.2.

$$f(S) = \sum_{X \subseteq S} (-1)^{|S \setminus X|} \cdot f\zeta(X) \tag{7.3}$$

Proof.

$$\sum_{X \subseteq S} (-1)^{|S \setminus X|} \cdot f\zeta(X) = \sum_{X \subseteq S} (-1)^{|S \setminus X|} \cdot \sum_{Y \subseteq X} f(Y)$$

$$= \sum_{Y \subseteq S} \left(\sum_{Y \subseteq X \subseteq S} (-1)^{|S \setminus X|} \right) \cdot f(Y)$$

Every nonempty set has equal number of even and odd-sized subsets. It means that
if $|S \setminus Y| = k > 0$, then the inner summand is equal to

$$\sum_{i=0}^{k} (-1)^i \binom{k}{i} = 0.$$

Thus

$$\sum_{Y \subseteq S} \left(\sum_{Y \subseteq X \subseteq S} (-1)^{|S \setminus X|} \right) \cdot f(Y) = f(S).$$

$\square$

We define the *Möbius transform* μ of a function $f : 2^{\mathcal{U}} \to \mathbb{Z}$ as the function
$f\mu : 2^{\mathcal{U}} \to \mathbb{Z}$ satisfying

$$f\mu(S) = \sum_{X \subseteq S} (-1)^{|S \setminus X|} f(X).$$

Exercise 7.3. By Lemma 7.2, we have that $f\zeta\mu = f$. We leave the proof that $f\mu\zeta = f$ as an exercise for the reader.

Summarizing, we have established the principle of Möbius inversion: the set $2^{\mathcal{U}}$
(and more generally, any finite partially ordered set) has a pair of mutually inverse

linear transformations, the zeta transform and the Möbius transform. We formalize it in the following lemma.

Lemma 7.4. *Zeta and Möbius transform are mutually inverse functions.*

For our algorithms we need to rank the functions discussed. The *ranked zeta transform* $f\zeta$ of a function $f : 2^{\mathcal{U}} \to \mathbb{Z}$ is defined for all $S \subseteq \mathcal{U}$ and all $k \in \{0, 1, \ldots, n\}$ as follows

$$f\zeta(k, S) = \sum_{X \subseteq S, |X| = k} f(X).$$

Let us remark that the zeta transform can be obtained in terms of the ranked zeta transform by taking the sum over k, i.e.

$$f\zeta(S) = \sum_{k=0}^{|S|} f\zeta(k, S).$$

Given the ranked zeta transform $f\zeta$, as in Lemma 7.2, one can show that the function f can be established by the following formula for inverting the ranked zeta transform.

$$f(S) = f\zeta(|S|, S) = \sum_{X \subseteq S} (-1)^{|S \setminus X|} \cdot f\zeta(|S|, X) \tag{7.4}$$

Now let us turn to algorithmic issues and study fast algorithms to compute ranked zeta and Möbius transforms. A cornerstone of fast subset convolution is the following lemma.

Lemma 7.5. *Let $M > 0$ be an integer, $\mathcal{U}$ be a set of size n and $f : 2^{\mathcal{U}} \to \mathbb{Z}$ be a function such that for every $S \subseteq \mathcal{U}$, $|f(S)| \leq M$. We also assume that all the values $f(S)$ for all the 2^n sets $S \subseteq \mathcal{U}$ can be computed in time $2^n \cdot \log M \cdot n^{\mathcal{O}(1)}$. Then the zeta and Möbius transforms can be computed in time $2^n \cdot \log M \cdot n^{\mathcal{O}(1)}$.*

Proof. For a function $f : 2^{\mathcal{U}} \to \mathbb{Z}$, we start by computing all values $f(S)$ for all $S \subseteq \mathcal{U}$, and keeping all these values in a table. By the assumption of the lemma, this takes time $2^n \cdot \log M \cdot n^{\mathcal{O}(1)}$, and we use space $2^n \cdot \log M \cdot n^{\mathcal{O}(1)}$ to keep this table. Once the table is computed, for every $S \subseteq \mathcal{U}$ we can look up the value of $f(S)$ in polynomial time.

The algorithm builds a table of entries $f\zeta_i(S)$ for all $S \subseteq \mathcal{U}$ and all $i \in \{0, 1, \ldots, n\}$ such that

$$f\zeta_i(S) = \sum_{S \setminus \{1, 2, \ldots, i\} \subseteq X \subseteq S} f(X).$$

This implies that $f\zeta_n(S) = f\zeta(S)$ for all $S \subseteq \mathcal{U}$, and thus we seek the entries $f\zeta_n(S)$. The dynamic programming algorithm works as follows. Initially, for all $S \subseteq \mathcal{U}$

$$f\zeta_0(S) = f(S).$$

Then iteratively for $i \geq 1$ the algorithm computes $f\zeta_i(S)$ for all $S \subseteq \mathcal{U}$ based on the following recurrence

$$f\zeta_i(S) = \begin{cases} f\zeta_{i-1}(S), & \text{if } i \notin S, \\ f\zeta_{i-1}(S \setminus \{i\}) + f\zeta_{i-1}(S), & \text{if } i \in S. \end{cases}$$

Each entry of the table requires at most $\log M$ bits. Thus the fast zeta transform $f\zeta$ of a given function f can be computed in such a way that all 2^n values of $f\zeta(X)$ are established in time $2^n \cdot \log M \cdot n^{\mathcal{O}(1)}$.

The fast Möbius transform $f\mu$ of a function $f : 2^{\mathcal{U}} \to \mathbb{Z}$ can be done by a similar dynamic programming algorithm. The algorithm builds a table of entries $f\mu_i(S)$ for all $S \subseteq \mathcal{U}$ and all $i \in \{0, 1, 2, \ldots, n\}$ such that

$$f\mu_i(S) = \sum_{S \setminus \{1,2,\ldots,i\} \subseteq X \subseteq S} (-1)^{|S \setminus X|} f(X).$$

This implies that $f\mu_n(S) = f\mu(S)$ for all $S \subseteq \mathcal{U}$. The dynamic programming algorithm works as follows. Initially, for all $S \subseteq \mathcal{U}$

$$f\mu_0(S) = f(S).$$

For $i \geq 1$ the algorithm computes $f\mu_i(S)$ for all $S \subseteq \mathcal{U}$ based on the following recurrence

$$f\mu_i(S) = \begin{cases} f\mu_{i-1}(S), & \text{if } i \notin S \\ -f\mu_{i-1}(S \setminus \{i\}) + f\mu_{i-1}(S), & \text{if } i \in S. \end{cases}$$

Thus the Möbius transform $f\mu$ can be computed in such a way that all the 2^n values $f\mu(S)$ are established within time $2^n \cdot \log M \cdot n^{\mathcal{O}(1)}$. $\qquad\square$

Corollary 7.6. *If $M = 2^{n^{\mathcal{O}(1)}}$ then the zeta and Möbius transforms can be computed in time $\mathcal{O}^*(2^n)$.*

7.2 Fast Subset Convolution

Now we present an algorithm for *fast subset convolution* which in principle is obtained by combining some dynamic programming algorithms. The fundamental idea is to use ranked zeta transforms and to define a convolution of ranked zeta transforms. Let $f\zeta$ and $g\zeta$ be two ranked zeta transforms. The ranked convolution of $f\zeta$ and $g\zeta$, denoted by $f\zeta \circledast g\zeta$, is defined for $k \in \{0, 1, 2, \ldots n\}$ and all $S \subseteq \mathcal{U}$ by

$$(f\zeta \circledast g\zeta)(k, S) = \sum_{j=0}^{k} f\zeta(j, S) \cdot g\zeta(k - j, S). \tag{7.5}$$

We emphasize that the *ranked convolution is over the rank parameter k*, while the *standard convolution is over the subset parameter T*. This definition of the ranked convolution and the definition of the ranked zeta transform (7.4) bring us to the following theorem.

Theorem 7.7. *Let $f, g : 2^{\mathcal{U}} \to \mathbb{Z}$ and let $f\zeta$ and $g\zeta$ be their ranked zeta transforms. Then $f * g$ is equal to the ranked Möbius transform of $f\zeta \circledast g\zeta$, i.e. for all $S \subseteq \mathcal{U}$:*

$$(f * g)(S) = \sum_{X \subseteq S} (-1)^{|S \setminus X|} (f\zeta \circledast g\zeta)(|S|, X) \tag{7.6}$$

Proof. To transform the right side of (7.6)

$$\sum_{X \subseteq S} (-1)^{|S \setminus X|} (f\zeta \circledast g\zeta)(|S|, X),$$

we first substitute the definition of the ranked convolution to obtain

$$\sum_{X \subseteq S} (-1)^{|S \setminus X|} \sum_{i=0}^{|S|} f\zeta(i, X) g\zeta(|S| - i, X).$$

The definition of the ranked zeta transform for $f\zeta(k, X)$ and $g\zeta(|S| - i, X)$ yields

$$\sum_{X \subseteq S} (-1)^{|S \setminus X|} \sum_{i=0}^{|S|} \Big(\sum_{\substack{Y \subseteq X \\ |Y| = i}} f(Y) \Big) \Big(\sum_{\substack{Z \subseteq X \\ |Z| = |S| - i}} g(Z) \Big)$$

$$= \sum_{X \subseteq S} (-1)^{|S \setminus X|} \sum_{i=0}^{|S|} \sum_{\substack{Y, Z \subseteq X \\ |Y| = i \\ |Z| = |S| - i}} f(Y) g(Z) \tag{7.7}$$

Now let us simplify (7.7) by determining the coefficient of $f(Y)g(Z)$ for all pairs (Y, Z). Clearly $f(Y)g(Z)$ occurs in the expression only if $|Y| + |Z| = |S|$. Furthermore, $f(Y)g(Z)$ occurs once for every X satisfying $Y \cup Z \subseteq X \subseteq S$ and its sign is then $(-1)^{|S \setminus X|}$. Hence in (7.7), the overall coefficient of $f(Y)g(Z)$, for any single (Y, Z) satisfying $|Y| + |Z| = |S|$, is equal to

$$\sum_{Y \cup Z \subseteq X \subseteq S} (-1)^{|S \setminus X|} = \sum_{i=0}^{|S \setminus (Y \cup Z)|} (-1)^i \binom{|S \setminus (Y \cup Z)|}{i}$$

$$= \begin{cases} 1, & \text{if } |S \setminus (Y \cup Z)| = 0 \text{ i.e. if } (Y, Z) \text{ is a partition of } S, \\ 0, & \text{otherwise.} \end{cases} \tag{7.8}$$

Combining (7.7) and (7.8), we obtain that for every $S \subseteq \mathcal{U}$

$$\sum_{X \subseteq S} (-1)^{|S \setminus X|} (f\zeta \circledast g\zeta)(|S|, X) = \sum_{\substack{Y, Z \subseteq S \\ Y \cup Z = S \\ Y \cap Z = \emptyset}} f(Y) g(Z)$$

By (7.2), we have that for all $S \subseteq \mathcal{U}$,

$$\sum_{X \subseteq S} (-1)^{|S \setminus X|} (f\zeta \circledast g\zeta)(|S|, X) = (f * g)(S),$$

and this completes the proof.   □

Given functions $f, g : 2^{\mathcal{U}} \to \mathbb{Z}$, based on Theorem 7.7 we can now evaluate $f * g$ as follows: First compute the fast ranked zeta transforms of f and g, then compute the ranked convolution of the zeta transforms $f\zeta$ and $g\zeta$, and finally invert the obtained result by using the fast ranked Möbius transform. Thus we arrive at the following theorem.

Theorem 7.8. *Let $\mathcal{U}$ be a set of size n and $f, g : 2^{\mathcal{U}} \to \mathbb{Z}$ be functions with range in $\{-M, -M+1, \ldots, M-1, M\}$, and such that all the values $f(X)$ and $g(X)$, for all $X \subseteq \mathcal{U}$, can together be computed in time $2^n \cdot \log M \cdot n^{\mathcal{O}(1)}$. Then the subset convolution $f * g$ can be computed in time $2^n \cdot \log M \cdot n^{\mathcal{O}(1)}$.*

Let us remark that in the proof of Theorem 7.8 and Lemma 7.5, we multiply $\mathcal{O}(n \log M)$-bit numbers. Thus to make our running time estimates more realistic, we may assume a model of computation, in which multiplication of two b-bit integers can be done in time $\mathcal{O}(b \log b \log \log b)$. This will lead to a running time of $2^n \cdot \log M \cdot (n \cdot \log M)^{\mathcal{O}(1)}$.

Many NP-hard combinatorial optimization problems do not deal with the integer sum-product ring to which Theorem 7.7 directly applies. However, this theorem can be adapted, for example, to compute for all $S \subseteq \mathcal{U}$ the following functions:

$$\max_{X \subseteq S} f(X) + g(S \setminus X),$$

and

$$\min_{X \subseteq S} f(X) + g(S \setminus X).$$

For example, to compute the above mentioned maximum, we can scale functions $f(X)$ and $g(X)$ and apply Theorem 7.8 to compute the subset convolution of functions $f_1(X) = \beta^{f(X)}$ and $g_1(X) = \beta^{g(X)}$, where $\beta = 2^n + 1$. Then the time required to compute $f_1 * g_1$ is $2^n \cdot M \cdot n^{\mathcal{O}(1)}$ Then to compute $\max_{X \subseteq S} f(X) + g(S \setminus X)$ the algorithm has to find a value p such that the result of the convolution is between β^{p-1} and β^p. All these computations can be done in time $2^n \cdot M \cdot n^{\mathcal{O}(1)}$.

Another remark is that all our proofs can be carried out on a much more general model. Instead of taking multiplication over integers, it is possible to prove similar results for functions with values in rings. Then instead of summation and multiplication over integers one can use multiplication and addition over a ring.

Let us mention that Theorem 7.8 can be proved by using the Fast Fourier Transform, for short FFT. Given two polynomials $F(x)$ and $G(x)$ of degree n, the naive way of computing the product $F(x) \cdot G(x)$ takes $\mathcal{O}(n^2)$ operations. By making use of FFT this can be speed up to $\mathcal{O}(n \log n)$. We refer to the corresponding chapter in Cormen et al. [52] for an introduction and further discussions of FFT. To compute the subset convolution $(f * g)(S)$ for all $S \subseteq \mathcal{U}$, we do the following.

Let us choose an (arbitrary) ordering $(u_1, u_2, \ldots, u_n)$ of the elements of $\mathcal{U}$. For a subset $S \subseteq \mathcal{U}$, we define

$$bin(S) = \sum_{i=0}^{n-1} f_S(u_{i+1}) 2^i,$$

where f_S is the characteristic function

$$f_S(u_i) = \begin{cases} 1, & \text{if } u_i \in S, \\ 0, & \text{otherwise.} \end{cases}$$

In other words, $bin(S)$ is the number obtained by transforming the characteristic vector of S from the binary to decimal numeral system.

For every j, $0 \leq j \leq n$, we define ranked polynomials

$$F^j(x) = \sum_{\substack{S \subseteq \mathcal{U} \\ |S| = j}} f(S) x^{bin(S)}$$

and

$$G^j(x) = \sum_{\substack{S \subseteq \mathcal{U} \\ |S| = j}} g(S) x^{bin(S)}.$$

For all values of j, the computation of $F^j(x)$ and $G^j(x)$ can be done in time $\mathcal{O}^*(2^n)$. The crucial observation is that for sets A, B such that $|A| + |B| = k$, $bin(A) + bin(B)$ has exactly k 1s in its binary representation if and only if $A \cap B = \emptyset$. For $k \geq 1$, we define $(F \circledast G)^k(x)$ as the polynomial obtained from the polynomial

$$\sum_{j=0}^{k} F^j(x) \cdot G^{k-j}(x)$$

by keeping only the summands of degree ℓ, where ℓ has exactly k 1s in its binary representation. Again, for each k, such computation can be done by making use of FFT in time $\mathcal{O}^*(2^n)$. Finally, we compute

$$H(x) = \sum_{j=0}^{n} (F \circledast G)^j(x).$$

For every $S \subseteq \mathcal{U}$, the value of $(f * g)(S)$ is the coefficient of $x^{bin(S)}$ in $H(x)$.

7.3 Applications and Variants

Fast subset convolution is a powerful tool for the design of fast exponential time algorithms. Nevertheless it is important to mention that algorithmic applications often require extensions or modifications of the standard approach presented here. This may require the use of auxiliary functions which facilitate convolution, and it may also require changes in the dynamic programming algorithms for the standard fast subset convolution. In the following sections we present examples illustrating the use of fast subset convolution.

In Chap. 4, inclusion-exclusion algorithms for generic covering and partition problems counting the number of k-coverings and k-partitions have been presented. This allowed the design and analysis of an $\mathcal{O}^*(2^n)$ time algorithm to compute the chromatic number of a graph.

All these results can also be achieved by an application of fast subset convolution. Let $\mathcal{U}$ be a set of n elements and let $\mathcal{S}$ be a family of subsets of $\mathcal{U}$. Recall that pairwise disjoint sets $S_1, S_2, \ldots, S_k$ form a k-partition of $\mathcal{U}$ into $\mathcal{S}$ if $S_i \in \mathcal{S}$, $1 \le i \le k$ and $S_1 \cup S_2 \cup \cdots \cup S_k = \mathcal{U}$. We always assume here that for every $S \subseteq \mathcal{U}$, there is a polynomial time algorithm checking whether $S \in \mathcal{S}$.

In Sect. 3.4 a simple dynamic programming algorithm counting the numbers of k-partitions in time $\mathcal{O}^*(3^n)$ is presented. Let us show how to use subset convolution to speed up this algorithm.

Theorem 7.9. *Let $\mathcal{U}$ be a set of n elements and let $\mathcal{S}$ be a family of subsets of $\mathcal{U}$. For any $k \ge 1$, the number of k-partitions of $\mathcal{U}$ into $\mathcal{S}$ is computable in time $\mathcal{O}^*(2^n)$.*

Proof. Let $f : 2^{\mathcal{U}} \to \{0, 1\}$ be an indicator function, i.e. for any $S \subseteq \mathcal{U}$, $f(S) = 1$ if and only if $S \in \mathcal{S}$. Then the number of k-partitions is

$$\sum_{\substack{Y_1, Y_2, \ldots, Y_k \subseteq \mathcal{U} \\ \cup_{i=1}^k Y_i = \mathcal{U} \\ Y_i \cap Y_j = \emptyset \text{ if } i \ne j}} \prod_{i=1}^k f(Y_i).$$

Let $f^{*k} : 2^{\mathcal{U}} \to \mathbb{Z}$ be a function defined as follows

$$f^{*k} = \underbrace{f * f * \cdots * f}_{k \text{ times}}.$$

By (7.2), for all $S \subseteq \mathcal{U}$

$$f^{*k}(S) = \sum_{\substack{Y_1, Y_2, \ldots, Y_k \subseteq S \\ \cup_{i=1}^k Y_i = S \\ Y_i \cap Y_j = \emptyset \text{ if } i \ne j}} \prod_{i=1}^k f(Y_i).$$

In particular, the number of k-partitions of $\mathcal{U}$ is equal to $f^{*k}(\mathcal{U})$. To compute $f^{*k}(\mathcal{U})$, for every $S \subseteq \mathcal{U}$ we compute $f^{*2}(S), f^{*3}(S), \ldots, f^{*k}(S)$ in $k-1$ subset convolutions.

Each of the steps uses fast subset convolution and is performed in time $\mathcal{O}^*(2^n)$. The number of steps can be reduced to $\mathcal{O}(\log k)$ by using the doubling trick—computing convolutions of $f^{*2^i}(S)$. $\qquad\square$

Theorem 7.9 is very general and many problems are special cases of it.

Of course, Theorem 7.9 yields a $\mathcal{O}^*(2^n)$ algorithm counting proper colorings of an n-vertex graph G, and even more generally, computing its chromatic polynomial. (We put $f(S) = 1$ if and only if S is an independent set.) We can also use it for natural extensions of the partitioning problem. For example, to count maximal k-colorable induced subgraphs of a graph G with vertex set V. Indeed, for $S \subseteq V$, $G[S]$ is k-colorable if and only if $f^{*k}(S) > 0$.

Domatic Number. A graph G has k-*domatic partition* $X_1, X_2, \ldots, X_k$ if every X_i is a dominating set of G. Theorem 7.9 can be used to count the number of k-domatic partitions in time $\mathcal{O}^*(2^n)$ by putting $f(S) = 1$ if and only if S is a dominating set.

Interesting applications of subset convolution can be achieved by replacing the indicator function in the proof of Theorem 7.9 by some other functions.

Counting Spanning Forests. A *spanning forest* of a graph G is an acyclic graph F spanning all vertices of G. In particular, if F is connected, then it is a *spanning tree*. The number of spanning trees in a graph is computable in polynomial time as the determinant of a maximal principal submatrix of the Laplacian of G, a classical result known as Kirchhoff's Matrix Tree Theorem. However, counting spanning forests is a #P-complete problem. For a set $S \subseteq V$, let $\tau(S)$ be the number of trees in the subgraph of G induced by S. Thus for every S, $\tau(S)$ is computable in polynomial time. The crucial observation here is that the number of spanning forests in G is equal to

$$\sum_{\substack{Y_1, Y_2, \ldots, Y_k \subseteq V \\ \cup_{i=1}^k Y_i = V \\ Y_i \cap Y_j = \emptyset \text{ if } i \neq j}} \prod_{i=1}^k \tau(Y_i).$$

Therefore, to find the number of forests in G with exactly k connected components, one has to compute τ^{*k} and this can be done in $\mathcal{O}^*(2^n)$.

Even more generally, in time $\mathcal{O}^*(2^n)$ one can compute the number of spanning subgraphs of G with k components and ℓ edges. For graph $G = (V, E)$, let $s_{k,\ell}(G)$ be the number of spanning subgraphs of G with k components and ℓ edges. To evaluate $s_{k,\ell}(G[X])$ for every $X \subseteq V$, one can use a two-part recurrence one rank $i \in \{0, 1, \ldots, n\}$ for $|X| = i$ at a time. Omitting the base cases, for $k \geq 2$ we have

$$s_{k,\ell}(G[X]) = \frac{1}{k} \sum_j \sum_{\emptyset \subsetneq Y \subsetneq X} s_{1,j}(G[Y]) s_{k-1,\ell-j}(G[X \setminus Y])$$

and for $k = 1$

$$s_{1,\ell}(G[X]) = \binom{m_X}{\ell} - \sum_{k \geq 2} s_{k,\ell}(G[X]),$$

where m_X is the number of edges in $G[X]$.

Tutte Polynomial. The computation of the Tutte polynomial can be seen as a generalization of graph coloring and counting spanning forests. The *Tutte polynomial* of a graph $G = (V, E)$ is the bivariate polynomial

$$T(G; x, y) = \sum_{A \subseteq E} (x - 1)^{r(E) - r(A)} (y - 1)^{|A| - r(A)},$$

where $r(A) = |V| - k(A)$ and $k(A)$ is the number of connected components of the graph induced in G by the edges of A.

Many graph parameters are points or regions of the so-called Tutte plane. For example,

- $T(G, 2, 1)$ is the number of spanning forests in G;
- $T(G, 1, 1)$ is the number of spanning trees in G;
- $T(G, 1, 2)$ is the number of connected subgraphs in G;
- $T(G, 2, 0)$ is the number of acyclic orientations of G;
- The chromatic polynomial $P(G, \lambda)$ of G is expressible as

$$P(G, \lambda) = (-1)^{r(E)} \lambda^{k(E)} T(G; 1 - \lambda, 0).$$

The computations of numbers $s_{k,\ell}(G)$, the number of spanning subgraphs with k components and ℓ edges, is important to compute the Tutte polynomial. The crucial fact (which we do not prove here) is that the Tutte polynomial can be expressed in the following form

$$T_G(x, y) = \sum_{k,\ell} s_{k,\ell}(G)(x - 1)^{k-c} (y - 1)^{\ell + k - n},$$

where G has c components and n vertices. The following theorem is due to Björklund, Husfeldt, Kaski, and Koivisto [27].

Theorem 7.10. *The Tutte polynomial of an n-vertex graph G can be computed in time $\mathcal{O}^*(2^n)$.*

In the remaining part of this section we present some variants that are relaxations of the subset convolution of form (7.2).

Let us recall that $\mathcal{U} = \{1, 2, \ldots, n\}$ is a set and $f, g : 2^{\mathcal{U}} \to \mathbb{Z}$ are two functions.

Definition 7.11. The *covering product* of f and g, denoted by $f *_c g$ is defined for all $S \subseteq \mathcal{U}$ as

$$(f *_c g)(S) = \sum_{\substack{X, Y \subseteq S \\ X \cup Y = S}} f(X) \cdot g(Y). \tag{7.9}$$

Definition 7.12. The *packing product* of f and g is defined for all $S \subseteq \mathcal{U}$ as

$$(f *_p g)(S) = \sum_{\substack{X, Y \subseteq S \\ X \cap Y = \emptyset}} f(X) \cdot g(Y). \tag{7.10}$$

Definition 7.13. The *intersecting covering product* of f and g, denoted by $f *_{ic} g$, is defined for all $S \subseteq \mathcal{U}$ as

$$(f *_{ic} g)(S) = \sum_{\substack{Y,Z \subseteq S \\ Y \cup Z = S \\ Y \cap Z \neq \emptyset}} f(Y) \cdot g(Z). \tag{7.11}$$

Now let us consider algorithms to compute these products.

Theorem 7.14. *Given functions* $f, g : 2^{\mathcal{U}} \to \mathbb{Z}$, *there is an algorithm to compute the covering product* $(f *_c g)(S)$ *for all* $S \subseteq \mathcal{U}$ *in time* $\mathcal{O}(n2^n)$.

Proof. Given f and g, the algorithm first computes the zeta transforms $f\zeta$ and $g\zeta$ in time $\mathcal{O}^*(2^n)$. Then taking the elementwise product of the transforms $(f\zeta \cdot g\zeta)(X) = f\zeta(X) \cdot g\zeta(X)$, the Möbius transform is applied to the result. Using fast zeta and Möbius transforms, as presented in Sect. 7.1, the algorithm needs time $\mathcal{O}^*(2^n)$.

To see that the algorithm indeed outputs $(f *_c g)(S)$, note that the result of the described Möbius transform is

$$(f\zeta \cdot g\zeta)\mu(S) = \sum_{X \subseteq S} (-1)^{|S \setminus X|} (f\zeta \cdot g\zeta)(X) = \sum_{X \subseteq S} (-1)^{|S \setminus X|} \sum_{Y,Z \subseteq X} f(Y)g(Z).$$

Now each ordered pair (Y, Z) of subsets of S, contributes $(-1)^{|S \setminus X|}$ to the sum in (7.3) for all X with $Y, Z \subseteq X$. Therefore the overall coefficient of $f(Y)g(Z)$ is 1 if $Y \cup Z = S$, otherwise the coefficient is 0. Consequently

$$\sum_{X \subseteq S} (-1)^{|S \setminus X|} (f\zeta \cdot g\zeta)(X) = \sum_{\substack{Y,Z \subseteq S \\ Y \cup Z = S}} f(Y) \cdot g(Z) = (f *_c g)(S).$$

$\square$

Given the algorithms for fast subset convolution and fast covering product, it is not hard to establish fast algorithms to compute the packing product and the intersecting covering product.

Theorem 7.15. *Given functions* $f, g : 2^{\mathcal{U}} \to \mathbb{Z}$, *there are algorithms to compute the packing product* $(f *_p g)(S)$ *and the intersecting covering product for all* $S \subseteq \mathcal{U}$ *in time* $\mathcal{O}^*(2^n)$.

Proof. Let f_1 be the function assigning to each subset of $\mathcal{U}$ the integer 1. Hence by (7.2), for all $h : 2^{\mathcal{U}} \to \mathbb{Z}$ and all $S \subseteq \mathcal{U}$

$$(h * f_1)(S) = \sum_{Y \subseteq S} h(Y).$$

Consequently, given functions f and g the algorithm computes

$$((f * g) * f_1)(S) = \sum_{\substack{Y,Z \subseteq S \\ Y \cap Z = \emptyset}} f(Y)g(Z) = (f *_p g)(S).$$

Thus the packing product $f *_p g$ can be evaluated by first computing the subset convolution $f * g$ and by then computing the subset convolution of $(f * g)$ and f_1. Using fast subset convolution the algorithm needs time $\mathcal{O}^*(2^n)$.

An algorithm to compute the intersecting covering product in time $\mathcal{O}^*(2^n)$ follows immediately from the observation

$$f *_{ic} g = f *_c g - f * g.$$

$\square$

7.4 f-width and Rank-width

Width parameters of graphs, which are often defined by graph decompositions, are of great importance in graph theory as well as in the design and analysis of graph algorithms. One of the best known width parameters of graphs is the treewidth to which Chap. 5 is devoted. Other well-known width parameters of graphs are clique-width and rank-width. In this section we consider a generalization of rank-width.

The f-width of a finite set $\mathcal{U}$ is a general framework for width parameters. Note that the f-width can be seen as the width parameter of a graph $G = (V, E)$. Though in general f-width is independent of the edge set E, various interesting width parameters for graphs can be obtained as the f-width of the vertex set V via the choice of f. If f is a cut-rank function then the f-width of V is the well-known graph parameter *rank-width*. Other width parameters of graphs that can be obtained as the f-width of V are the *carving-width* and the *branching-width of a connectivity function* f. In this section we present a fast subset convolution based $\mathcal{O}^*(2^n)$ time algorithm to compute the f-width of a graph, assuming that f is given by an oracle and that the integer values of f are small.

f-width of Sets. Let $\mathcal{U}$ be a set of n elements and let $f : 2^{\mathcal{U}} \to \mathbb{Z}$ be a function assigning to each subset of $\mathcal{U}$ an integer. A *rooted binary tree* T is a directed tree with a specified vertex r called the root such that the root r has two incoming edges and no outgoing edges and every vertex other than the root has exactly one outgoing edge and either two or zero incoming edges. A leaf of a rooted binary tree is a vertex with no incoming edges. A descendent of an edge e of a rooted binary tree T is the set of vertices from which there exists a directed path to e.

A *decomposition* of a set $\mathcal{U}$ is a pair (T, μ) of a rooted binary tree T and a bijection μ from $\mathcal{U}$ to the set of all leaves of T. For a decomposition (T, μ) of $\mathcal{U}$ and an edge e of the binary rooted tree T, let $X_e \subseteq \mathcal{U}$ be the set of all elements of $\mathcal{U}$ being assigned to a leaf of T which is also a descendent of e. Now we define the *f-width of a decomposition* (T, μ) of $\mathcal{U}$ as the minimum of $f(\mu^{-1}(X_e))$ over all edges e of T. Finally, the *f-width* of a finite set $\mathcal{U}$, denoted by $w_f(\mathcal{U})$, is the minimum f-width over all possible decompositions of $\mathcal{U}$. If $|\mathcal{U}| \le 1$ then $\mathcal{U}$ has no decomposition but we let $w_f(\mathcal{U}) = f(\mathcal{U})$.

Let $G = (V,E)$ be a graph. There are three important width parameters of G that can be defined via f-width.

Branch-width. For branch-width, we put $\mathcal{U} = E$, the edge set of G. For every $X \subseteq E$ its *border* $\delta(X)$ is the set of vertices from V such that every $v \in \delta(X)$ is adjacent to an edge from X and to an edge from $E \setminus X$. The *branch-width* of G is the f-width of $\mathcal{U} = E$ with $f(X) = |\delta(X)|$. This definition can be extended to hypergraphs and matroids.

Carving-width. The *carving-width* of a graph $G = (V,E)$ is the f-width of $\mathcal{U} = V$, with $f(X) = CUT(X, V \setminus X)$, the number of edges between X and $V \setminus X$.

Rank-width. The *rank-width* of a graph $G = (V,E)$ is the f-width of $\mathcal{U} = V$, where $f = \rho_G$, the cut-rank function of G. Here the cut-rank function is defined as follows. For a vertex subset $X \subseteq V$, let

$$B_G(X) = (b_{i,j})_{i \in X, j \in V \setminus X},$$

be the $|X| \times |V \setminus X|$ matrix over the binary field GF(2) such that $b_{i,j} = 1$ if and only if $\{i,j\} \in E$. In other words, $B_G(X)$ is the adjacency matrix of the bipartite graph formed from G by removing all edges but the edges between X and $V \setminus X$. Finally, $\rho_G(X)$ is the rank of $B_G(X)$.

The following recursive definition of w_f will be used to compute the f-width $w_f(\mathcal{U})$.

Lemma 7.16. *Let $\mathcal{U}$ be a finite set and let $f : 2^{\mathcal{U}} \rightarrow \mathbb{Z}$ be a function. Then for all nonempty subsets $X \subseteq \mathcal{U}$,*

$$w_f(X) = \begin{cases} \min_{\emptyset \subsetneq Y \subsetneq X} \max\left(f(Y), f(X \setminus Y), w_f(Y), w_f(X \setminus Y)\right) & \text{if } |X| \geq 2 \\ f(X) & \text{if } |X| = 1 \end{cases} \quad (7.12)$$

Lemma 7.16 allows a simple dynamic programming exponential time algorithm to compute the f-width of a finite set $\mathcal{U}$ based on the recurrence (7.12).

Let us assume that $n = |\mathcal{U}|$ and $M = \max_{X \subseteq \mathcal{U}} |f(X)|$. Furthermore we assume that an oracle computes the integer value of any $f(X)$ in time $n^{O(1)}$. Under these assumptions the simple dynamic programming algorithm to compute $w_f(X)$ for all $X \subseteq \mathcal{U}$ needs time $\mathcal{O}^*(2^n)$ to compute $f(X)$ for all $X \subseteq \mathcal{U}$ using the oracle and then time $\sum_{k=2}^{n} \binom{n}{k} 2^k \log M n^{O(1)}$ to compute $w_f(X)$ for all $X \subseteq \mathcal{U}$. Hence the overall running time is $3^n \log M n^{O(1)}$.

Is it possible to compute $w_f(X)$ for all $X \subseteq \mathcal{U}$ faster, say in time proportional to 2^n? Or at least, can $w_f(\mathcal{U})$ be computed faster, which is the value of interest in case of rank-width of graphs, carving-width of graphs, etc.

Here we can observe some typical indications that fast subset convolution should be considered. There is a 3^n factor in the running time of the algorithm when implementing it directly. One may hope to replace it by 2^n via fast subset convolution. The algorithm deals with subsets of sets and it computes $w_f(X)$ for all subsets of $\mathcal{U}$.

Unfortunately when inspecting (7.12), we observe that the crucial part of the computation, the one that should be done by fast subset convolution, is

$$\min_{\emptyset \subseteq Y \subseteq X} \max(f(Y), f(X \setminus Y), w_f(Y), w_f(X \setminus Y)).$$

The following theorem shows how to avoid this obstacle.

Theorem 7.17. *There is an $2^n \log^2 M n^{O(1)}$ algorithm to compute the f-width of a finite set $\mathcal{U}$.*

Proof. We construct auxiliary functions in order to apply the fast subset convolution. We start by computing and keeping in a table all values $f(X)$, $X \subseteq \mathcal{U}$. This takes time $2^n \log M n^{O(1)}$.

We first design an algorithm that decides for any fixed value k whether $w_f(X) \leq k$ for all $X \subseteq \mathcal{U}$, assuming that $f(X)$ for all $X \subseteq \mathcal{U}$ is given in a table. First we transform this into a binary table where the entry for X is 1 if and only if $f(X) \leq k$. We define functions g_i for all $i \in \{1, 2, \ldots, n\}$ and for all $X \subseteq \mathcal{U}$ as follows:

$$g_i(X) = \begin{cases} 1, & \text{if } 1 \leq |X| \leq i, f(X) \leq k, w_f(X) \leq k, X \neq \mathcal{U}, \\ 1, & i = n, X = \mathcal{U}, w_f(X) \leq k, \\ 0, & \text{otherwise.} \end{cases} \qquad (7.13)$$

Hence by Lemma 7.16,

$$w_f(X) \leq k \text{ if and only if } \begin{cases} (g_{|X|-1} * g_{|X|-1})(X) \neq 0, & \text{if } |X| \geq 2, \\ f(X) \leq k, & \text{if } |X| = 1. \end{cases} \qquad (7.14)$$

By (7.14), the algorithm recursively computes g_{i+1} from g_i for $i \in \{1, 2, \ldots, n\}$ as follows. Constructing g_1 is easy. By Theorem 7.8, $g_i * g_i$ can be computed by fast subset convolution in time $\mathcal{O}^*(2^n \log M)$.

Finally, to compute $w_f(X)$ for all $X \subseteq \mathcal{U}$ we need to run the above algorithm for all possible values of k, i.e. M times. Should we only need to compute $w_f(\mathcal{U})$ then binary search will do and thus we have to run the above algorithm $\log M$ times. $\qquad \square$

This has the following consequences.

Corollary 7.18. *Let G be a graph with n vertices and m edges. The rank-width and the carving-width of G can be computed in time $\mathcal{O}^*(2^n)$. The branch-width of G can be computed in time $\mathcal{O}^*(2^m)$.*

Notes

The history of fast transforms can be traced back to the work of Yates [222]. We refer to Sect. 4.6 of Knuths's book [135] for more information on Yates's algorithm and fast Fourier transform. The influential paper of Rota advanced the general theory of Möbius inversions on posets [192].

The algorithm for fast subset convolution is due to Björklund, Husfeldt, Kaski, and Koivisto [26]. The proofs in [26] are given in a more general setting for rings and semirings. FFT is discussed in many textbooks on algorithms, see, e.g. [52]. The proof of Theorem 7.8 by making use of FFT was brought to our attention by K. Venkata and Saket Saurabh. Exact algorithms for the DOMATIC NUMBER problem were given by several authors [89, 180]. The $\mathcal{O}^*(2^n)$ algorithm for DOMATIC NUMBER is due to Björklund, Husfeldt, and Koivisto [30]. The algorithm evaluating the Tutte polynomial is due to Björklund, Husfeldt, Kaski, and Koivisto [27].

The results of Sect. 7.4 are based on the work of Oum [167]. Branch-width was defined by Robertson and Seymour in their fundamental work on Graph Minors [184]. The rank-width was defined by Oum and Seymour in [168]. Surveys [94, 113] provide more information on these parameters. The existence of an $\mathcal{O}^*(2^n)$ algorithm computing the branch-width of an n-vertex graph is an open problem. The best known vertex-exponential algorithm for this problem runs in time $\mathcal{O}^*((2\sqrt{3})^n)$ [93].

Lokshtanov and Nederlof develop a general framework of transforming dynamic programming algorithms to algorithms using only polynomial space [154]. In particular, their approach provides polynomial space and pseudo-polynomial time algorithms for SUBSET SUM and KNAPSACK.

Subset convolution can also be used in dynamic programming algorithms for graphs of bounded treewidth. For example, van Rooij, Bodlaender, and Rossmanith [189] used subset convolution to show that the number of perfect matchings can be counted in time $\mathcal{O}(2^t n)$ for graphs of treewidth at most t.

Chapter 8
Local Search and SAT

In Chap. 2, we discuss a branching algorithm for the k-SATISFIABILITY problem. In this chapter we consider more techniques for solving k-SAT. Both techniques are based on performing local search in balls in the Hamming space around some assignments. The first algorithm randomly chooses an assignment and performs a random walk of short length (in Hamming distance) to search for the solution. The second algorithm is deterministic and uses a similar idea; but instead of using a random walk, it finds a covering of the Hamming space by balls of specified radius and performs a search inside these balls.

The algorithms in this chapter are heavily based on the notions of Hamming space and Hamming distance.

Hamming space (of dimension n) is the set of all 2^n binary strings of length n. The *Hamming distance* between two binary strings is the number of positions in which the corresponding symbols of the strings are different. For example, the distance between the strings $(\mathbf{0},1,1,\mathbf{1},0)$ and $(\mathbf{1},1,1,\mathbf{0},0)$ is 2. Clearly the distance between two strings of length n is always at most n.

It is convenient to view satisfying assignments of CNF formulae as elements of the Hamming space. For a given CNF formula with n variables, we fix (arbitrarily) an ordering of the variables. Then every satisfying assignment corresponds to a binary string of length n. The value of the ith variable in the satisfying assignment is true if and only if the element in the ith position of the string is 1. Furthermore, we also will view the set of all truth assignments as the Hamming space $\mathcal{H}_n = \{0,1\}^n$. The *Hamming distance* between two truth assignments a and b is the number of bits on which the two assignments differ. In other words, this is the number of bits one has to flip to obtain b from a. For a truth assignment a and an integer d, we denote by $\mathcal{H}(a,d)$ the *ball of radius d centered in a*, which is the set of all truth assignments at Hamming distance at most d from assignment a. The number of assignments at distance exactly i from assignment a is $\binom{n}{i}$ and the volume of the ball, which is the cardinality of $\mathcal{H}(a,d)$, is

$$|\mathcal{H}(a,d)| = \sum_{i=0}^{d} \binom{n}{i}.$$

F.V. Fomin, D. Kratsch, *Exact Exponential Algorithms*, Texts in Theoretical
Computer Science. An EATCS Series, DOI 10.1007/978-3-642-16533-7_8,
© Springer-Verlag Berlin Heidelberg 2010

8.1 Random Walks to Satisfying Assignments

In this section we present a randomized algorithm solving the k-SAT problem. Its analysis is based on the following lemma.

Lemma 8.1. *Let F be a satisfiable formula with n variables. Assume that algorithm* $k-sat3$ *described in Fig. 8.1 runs on F until it finds a satisfying assignment. Then the expected number of steps of the algorithm is* $\mathcal{O}(n^{3/2} \left(\frac{2(k-1)}{k} \right)^n)$.

Proof. Let a^* be a satisfying assignment of F. For $j \in \{0, \ldots, n\}$, let q_j be the probability that the procedure $random-walk$ described in Fig. 8.2 finds a^* when starting from an assignment which is at distance j from a^*. Clearly, $q_0 = 1$. When we choose an unsatisfied clause and flip the value of one of its variables, we obtain a new assignment a'. The Hamming distance from $a*$ to the new assignment decreases by one with probability at least $1/k$ and increases by one with probability at most $(k-1)/k$.

Algorithm k-sat3(F).
Input: A k-CNF formula F with n variables and m clauses.
Output: A satisfying assignment of F.

repeat
    choose an assignment a of F uniformly at random
    $random-walk(a)$
until a *is satisfying*
return a

Fig. 8.1 Algorithm $k-sat3$ for k-SATISFIABILITY

Procedure random-walk.
Input: A truth assignment a of a k-CNF formula F with n variables.
Output: An assignment of F.

$count := 0$
$b := a$
while b is not a satisfying assignment and $count < 3n$ **do**
    choose an arbitrary clause which is not satisfied by b
    choose one variable of this clause uniformly at random and flip its value in the
    assignment b
    $count := count + 1$
return assignment b

Fig. 8.2 Procedure $random-walk$

To find a lower bound on q_j, we model the process as a random walk of a particle on a path with vertices labelled by integers $0, 1, \ldots, n$. We assume that a particle moves from left to right (from vertex $j-1$ to j) with probability $(k-1)/k$ and from right to left (from j to $j-1$) with probability $1/k$. Then q_j is at least the probability of the event that a particle reaches 0 in at most $3n$ moves starting from j. (This bound is a pessimistic view of the process because for some cases the probability of decreasing the distance to a^* can be larger than $1/k$.)

To give a lower bound on q_j, we estimate the probability of reaching 0 in $j+2\ell \le 3n$ steps if $j+\ell$ steps are the moves from right to left and ℓ steps are from left to right. The probability of this event is

$$\binom{j+2\ell}{\ell}\left(\frac{k-1}{k}\right)^{\ell}\left(\frac{1}{k}\right)^{j+\ell}.$$

Thus

$$q_j \ge \max_{0\le\ell\le(3n-j)/2}\binom{j+2\ell}{\ell}\left(\frac{k-1}{k}\right)^{\ell}\cdot\left(\frac{1}{k}\right)^{j+\ell}.$$

Using the binary entropy function and Lemma 3.13, we have that

$$\binom{(1+2\alpha)j}{\alpha j} \ge \frac{1}{\sqrt{8j\alpha(1-\alpha)}}\cdot\left[\left(\frac{1+2\alpha}{\alpha}\right)^{\alpha}\left(\frac{1+2\alpha}{1+\alpha}\right)^{1+\alpha}\right]^{j}.$$

By putting $\alpha = \frac{1}{k-2}$, we arrive at

$$q_j \ge \frac{1}{\sqrt{8j}}\cdot\left[\left(\frac{1+2\alpha}{\alpha}\right)^{\alpha}\cdot\left(\frac{1+2\alpha}{1+\alpha}\right)^{1+\alpha}\cdot\left(\frac{k-1}{k}\right)^{\alpha}\cdot\left(\frac{1}{k}\right)^{1+\alpha}\right]^{j}$$

$$\ge \frac{1}{\sqrt{8j}}\left(\frac{1}{k-1}\right)^{j}.$$

Let p_j, $0 \le j \le n$, be the probability of choosing at random an assignment at distance j from a^*. Then $p_0 = \frac{1}{2^n}$, and for $j \ge 1$,

$$p_j = \binom{n}{j}\left(\frac{1}{2}\right)^{n}.$$

Let q be the probability that the procedure random-walk finds a^* (or some other satisfying assignment). We have that

$$q \geq \sum_{j=0}^{n} p_j q_j \geq \left(\frac{1}{2}\right)^n + \sum_{j=1}^{n} p_j q_j$$

$$\geq \left(\frac{1}{2}\right)^n + \sum_{j=1}^{n} \binom{n}{j} \left(\frac{1}{2}\right)^n \frac{1}{\sqrt{8j}} \left(\frac{1}{k-1}\right)^j$$

$$\geq \frac{1}{\sqrt{8n}} \left(\frac{1}{2}\right)^n \sum_{j=1}^{n} \binom{n}{j} \left(\frac{1}{k-1}\right)^j$$

$$= \frac{1}{\sqrt{8n}} \left(\frac{1}{2}\right)^n \left(1 + \frac{1}{k-1}\right)^n = \frac{1}{\sqrt{8n}} \left(\frac{k}{2(k-1)}\right)^n.$$

Now we use the following well-known result.

Lemma 8.2. *When we repeatedly perform independent trials of an experiment, each of which succeeds with probability p, then the expected number of steps we need to perform until the first success is $1/p$.*

Proof. Let X be the random variable equal to the number of trials performed until the first success occurs. Then the probability that $X = i$ is equal to $p(1-p)^{i-1}$, and we have that

$$E[X] = \sum_{i=0}^{\infty} i \cdot \Pr[X = i] = \sum_{i=0}^{\infty} ip(1-p)^{i-1} = -p\frac{d}{dp} \sum_{i=0}^{\infty} (1-p)^i$$

$$= -p\frac{d}{dp}\frac{1}{p} = -p\frac{-1}{p^2} = \frac{1}{p}.$$

$\square$

Thus the expected number of random assignments for which algorithm $k-\text{sat}3$ calls procedure random-walk is at most $1/q$. The running time of Random-walk is $\mathcal{O}(n)$, and thus the expected running time of $k-\text{sat}3$ is $\mathcal{O}(n^{3/2} \left(\frac{2(k-1)}{k}\right)^n)$. $\square$

Lemma 8.1 implies the following Monte Carlo algorithm for k-SAT.

Theorem 8.3. *Let F be a k-CNF formula with n variables. Then there is an algorithm for k-SAT with the following properties. If F is not satisfiable, then the algorithm returns the correct answer that F is unsatisfiable. If F is satisfiable, then for every integer $b > 0$, with probability at least $1 - 2^{-b}$ the algorithm returns a satisfying assignment after $\mathcal{O}(b \cdot n^{3/2} \cdot \left(2 - \frac{2}{k}\right)^n)$ steps. Otherwise, the algorithm incorrectly reports that F is unsatisfiable.*

Proof. Let us consider the algorithm $k-\text{sat}4$ in Fig. 8.3. If F does not have a satisfying assignment, then the algorithm does not find a satisfying assignment and correctly concludes that the formula is unsatisfiable.

Let us assume that F is satisfiable and let $b > 0$ be an integer. We run the algorithm by selecting

$$2b \cdot \left\lceil \sqrt{8n} \left(\frac{2(k-1)}{k}\right)^n \right\rceil$$

Algorithm k−sat4(F).
Input: A k-CNF formula F with n variables and m clauses.
Output: Either a satisfying assignment of F or a report that F is unsatisfiable.

$count := 0$
repeat
$\quad\Big|\quad$ choose an assignment a of F uniformly at random
$\quad\Big|\quad$ random-walk(a)
$\quad\Big|\quad$ $count := count + 1$
until a is a satisfying assignment or $count \geq 2b \cdot \left\lceil \sqrt{8n} \left(\frac{2(k-1)}{k} \right)^n \right\rceil$

if a satisfies F **then**
$\quad\Big|\quad$ **return** a
else
$\quad\quad$ **return** F is unsatisfiable

Fig. 8.3 Algorithm k-sat4 for k-SATISFIABILITY

---

times a random truth assignment a and running for each such assignment the procedure random-walk. Thus the total running time of the algorithm is $\mathcal{O}(b \cdot n^{3/2} \cdot \left(\frac{2(k-1)}{k} \right)^n)$. To analyse the algorithm, we partition its execution into b segments; each of the segments consists of

$$2 \cdot \left\lceil \sqrt{8n} \left(\frac{2(k-1)}{k} \right)^n \right\rceil$$

calls of random-walk.

Let X_i be a random variable equal to the number of times procedure random-walk is called from the start of the ith segment until we find a satisfying assignment. In Lemma 8.1, the estimation of the probability of reaching a satisfying assignment q does not depend on the starting position. Thus, we have that

$$E(X_i) \leq \sqrt{8n} \left(\frac{2(k-1)}{k} \right)^n.$$

By Markov's inequality, for any $c > 0$,

$$\Pr[X_i > c] \leq \frac{E(X_i)}{c}.$$

Thus

$$\Pr\left[X_i > 2 \left\lceil \sqrt{8n} \left(\frac{2(k-1)}{k} \right)^n \right\rceil\right] \leq \frac{\sqrt{8n} \left(\frac{2(k-1)}{k} \right)^n}{2 \left\lceil \sqrt{8n} \left(\frac{2(k-1)}{k} \right)^n \right\rceil} \leq \frac{1}{2}.$$

In other words, the probability that the algorithm does not find a satisfying assignment during the execution of the ith segment is at most $1/2$. Thus the probability that the algorithm fails in all b segments is at most 2^{-b}.

$\square$

8.2 Searching Balls and Cover Codes

In this section we describe a deterministic algorithm solving k-SAT which has running time $\mathcal{O}^*\left(\left(\frac{2k}{k+1}\right)^n\right)$. In previous section, we chose balls at random and then searched balls for a satisfying assignment. In this section the strategy is to (deterministically) cover the Hamming space of all assignments by a sufficiently small amount of balls of small radius.

The following lemma gives a deterministic algorithm searching a ball of radius r.

Lemma 8.4. *Given a k-CNF formula F and a truth assignment a of F, it can be decided in time $k^r n^{\mathcal{O}(1)}$ whether there is a satisfying assignment of F which is at Hamming distance at most r from the assignment a.*

Proof. If a is not a satisfying assignment, the algorithm chooses an arbitrary unsatisfied clause C of F. Then it generates at most k new assignments by flipping the value of each variable in C. Each of the new assignments is at Hamming distance one from a. Moreover, if there is a satisfying assignment a^*, then the distance from at least one new assignment to a^* is smaller than the distance from a to a^*. The algorithm proceeds recursively. At every step of the recursion it creates at most k new assignments and the depth of the recursion is at most r. Thus in the search tree corresponding to an execution of the branching algorithm all interior nodes have at most k children and its height is at most r. Consequently the search tree has at most k^r leaves and thus the running time is $\mathcal{O}^*(k^r)$. Consequently the algorithm finds in time $\mathcal{O}^*(k^r)$ either a satisfying assignment, or it correctly concludes that there is no satisfying assignment within Hamming distance at most r from a. $\qquad\square$

Let 0^n be the truth assignment with all variables set to 0 (false) and 1^n that with all variables set to 1 (true). Then every assignment, including a satisfying assignment, belongs to one of the balls $\mathcal{H}(0^n, n/2)$ and $\mathcal{H}(1^n, n/2)$. Then by Lemma 8.4, we can (deterministically) solve k-SAT in time $(\sqrt{k})^n \cdot n^{\mathcal{O}(1)}$. While for $k = 3$ this is better than the brute force search, for $k \geq 4$ the result is not so exciting. What happens if we try to use balls of smaller sizes?

Let us study the following randomized algorithm. Choose uniformly at random a truth assignment a and search for a satisfying assignment at distance at most αn from a. The probability that a satisfying assignment is in $\mathcal{H}(a, \alpha n)$ is at least

$$\frac{\sum_{i=1}^{\alpha n} \binom{n}{i}}{2^n}$$

(the ratio of $|\mathcal{H}(a, \alpha n)|$ and the total number of assignments). The expected number of steps the algorithm runs until it finds a satisfying assignment, assuming there is one, is

$$\mathcal{O}^*\left(\frac{2^n}{\sum_{i=1}^{\alpha n} \binom{n}{i}} \cdot k^{\alpha n}\right) = \mathcal{O}^*(2^{(1-h(\alpha))n} \cdot k^{\alpha n}),$$

where $h(\cdot)$ is the binary entropy function. Straightforward differentiation shows that we can minimize the exponential part of the product by choosing $\alpha = \frac{1}{k+1}$. Thus the expected number steps of the algorithm is

$$\mathcal{O}^* \left(\left(\frac{2k}{k+1} \right)^n \right).$$

Now, as in the proof of Theorem 8.3, it is easy to obtain a Monte Carlo algorithm solving k-SAT which has running time $\mathcal{O}^* \left(\left(\frac{2k}{k+1} \right)^n \right)$.

In what follows, we show how to make this algorithm deterministic without changing its exponential running time.

A *code of length* n is a subset of $\mathcal{H}_n$. A *covering code* C *of radius* d is a code such that every truth assignment is within distance at most d from some assignment in C. We start with the bounds on the size of a covering code.

What is the size of a covering code of radius αn? The following lemma answers this question.

Lemma 8.5. *Let* C *be a covering code of radius* αn, $0 < \alpha \leq 1/2$. *Then*

$$|C| \geq \frac{2^n}{|\mathcal{H}(0^n, \alpha n)|} \geq 2^{(1-h(\alpha))n}.$$

Proof. Every ball of radius αn covers at most $|\mathcal{H}(0^n, \alpha n)|$ truth assignments. Thus to cover 2^n assignments, one needs at least

$$\frac{2^n}{|\mathcal{H}(0^n, \alpha n)|}$$

balls.   □

The bound of Lemma 8.5 is tight up to polynomial factor.

Lemma 8.6. *For every* $0 < \alpha \leq 1/2$ *there exists a covering code* C *of radius* αn *such that*

$$|C| \leq \left\lceil n \frac{2^n}{|\mathcal{H}(0^n, \alpha n)|} \right\rceil \leq n \sqrt{n\alpha(1-\alpha)} \cdot 2^{(1-h(\alpha))n}.$$

Proof. Let us choose uniformly at random $\left\lceil n \frac{2^n}{|\mathcal{H}(0^n, \alpha n)|} \right\rceil$ assignments (with possible repetitions). We want to show that with a positive probability this is a covering code.

Let a be an assignment. The probability that the first chosen element is at distance more than αn from a, or is not in $\mathcal{H}(a, \alpha n)$ is

$$1 - \frac{\mathcal{H}(a, \alpha n)}{2^n} = 1 - \frac{\mathcal{H}(0^n, \alpha n)}{2^n}.$$

Thus the probability that a is not covered by $\left\lceil n \frac{2^n}{|\mathcal{H}(0^n, \alpha n)|} \right\rceil$ assignments is

$$\left(1 - \frac{\mathcal{H}(0^n, \alpha n)}{2^n}\right)^{\left\lceil n \frac{2^n}{|\mathcal{H}(0^n, \alpha n)|}\right\rceil} \le e^{-n},$$

where $e = \lim_{n \to +\infty}(1 + \frac{1}{n})^n \sim 2.718282$.

Therefore, the probability that there is a non-covered assignment is at most $2^n e^{-n}$, and we conclude that the probability that the chosen assignment is a covering code is at least

$$1 - \frac{2}{e} > 0.$$

We have shown that a randomly chosen assignment set of size $\left\lceil n \frac{2^n}{|\mathcal{H}(0^n, \alpha n)|}\right\rceil$ with positive probability covers all assignments, which yields that a covering code of such cardinality exists.   □

Lemma 8.6 does not provide an algorithm to compute a covering code of given radius. This can be done by treating the problem as a set cover problem and running a greedy algorithm to find a set cover.

Lemma 8.7. *For every $0 < \alpha \le 1/2$ a covering code $\mathcal{C}$ of radius αn of size at most*

$$n^2 \sqrt{n\alpha(1-\alpha)} \cdot 2^{(1-h(\alpha))n}$$

can be computed in time $\mathcal{O}^*(8^n)$.

Proof. It is well known that a greedy algorithm for the MINIMUM SET COVER problem (the algorithm which always selects a set which covers the maximum number of yet uncovered elements) is a $\log M$-approximation algorithm, where M is the number of sets. In our case, we can view the covering code as an instance of the MINIMUM SET COVER problem with 2^n sets, each consisting of a ball of radius αn centered in some truth assignment. Thus the greedy approach provides a solution within a factor n of the optimal one. By Lemma 8.6, the size of the solution found is at most

$$n^2 \sqrt{n\alpha(1-\alpha)} \cdot 2^{(1-h(\alpha))n}.$$

To implement the greedy approach, we keep for every ball the set of assignments already covered by the selected balls. Every time we select a new ball, we recompute these sets. The number of such selections is at most 2^n and for every selection a recomputation of covered assignments takes time $\mathcal{O}^*(4^n)$. Thus the total running time is $\mathcal{O}^*(8^n)$.   □

The following lemma trades off the size of the covering code and the time required for its computation.

Lemma 8.8. *Let $\ell \ge 2$ be a divisor of n. For every $0 < \alpha \le 1/2$ a covering code $\mathcal{C}$ of radius αn of size at most*

$$\left(n^2 \sqrt{n\alpha(1-\alpha)}\right)^\ell \cdot 2^{(1-h(\alpha))n}$$

can be computed in time $(2^{3n/\ell} + 2^{(1-h(\alpha))n}) \cdot n^{\mathcal{O}(1)}$.

Proof. We partition the set of n variables into ℓ groups of size n/ℓ. All possible truth assignments for each of these groups form a Hamming space $\mathcal{H}_{n/\ell}$. Every truth assignment in $\mathcal{H}_n$ is a concatenation of ℓ assignments in $\mathcal{H}_{n/\ell}$. Moreover, every covering code in $\mathcal{H}_n$ with radius αn is a concatenation of covering codes in $\mathcal{H}_{n/\ell}$ with radius $\alpha n/\ell$.

By Lemma 8.7, a covering code of radius $\alpha n/\ell$ in $\mathcal{H}_{n/\ell}$ of size at most

$$n^2 \sqrt{n\alpha(1-\alpha)} \cdot 2^{\frac{(1-h(\alpha))n}{\ell}}$$

can be computed in time $2^{3n/\ell} \cdot n^{\mathcal{O}(1)}$.

By trying all possible concatenations of codes obtained for $\mathcal{H}_{n/\ell}$, we obtain the code for $\mathcal{H}_n$ and the size of this code is at most

$$\left(n^2 \sqrt{n\alpha(1-\alpha)}\right)^{\ell} \cdot 2^{(1-h(\alpha))n}.$$

$\square$

We are ready to prove the main result of this section.

Theorem 8.9. *The deterministic algorithm* k-sat 5 *solves the* k-SAT *problem with* n *variables in time*

$$\mathcal{O}^*\!\left(\left(2 - \frac{2}{k+1}\right)^n\right).$$

Algorithm k–sat5(F).
Input: A k-CNF formula F with n variables and m clauses.
Output: Either a satisfying assignment of F or a report that F is unsatisfiable.

$\alpha = 1/(k+1)$
use Lemma 8.8 with $\ell = 6$ to construct a covering code $\mathcal{C}$ of length n and radius αn
forall assignments $a \in \mathcal{C}$ **do**
    $\lfloor$ use Lemma 8.4 to search for a satisfying assignment in the ball $\mathcal{H}(a, \alpha n)$

if satisfying assignment a is found **then**
    | **return** a
else
    $\lfloor$ **return** F is unsatisfiable

Fig. 8.4 Algorithm k–sat5 for k-SATISFIABILITY

Proof. Algorithm k–sat5 given in Fig. 8.4 is based on local search. It first constructs a covering code $\mathcal{C}$ and then it searches all balls around the assignments of the code for a satisfying assignment. Without loss of generality we assume here that n is divisible by 6, otherwise we can add at most 5 fake variables, increasing the complexity by a polynomial factor. Because $\mathcal{C}$ is a covering code, clearly, the algorithm either finds a satisfying assignment or correctly reports that the formula is not satisfiable.

The running time of the algorithm is the sum of the following running times: the time $T_1(n, \alpha n, \ell)$ needed to construct $\mathcal{C}$ of radius αn plus $|\mathcal{C}| \cdot T_2(n, \alpha n)$, where $T_2(n, \alpha n)$ is the time required to search a ball of radius αn.

For $\alpha = 1/(k+1)$ and $\ell = 6$, by Lemma 8.8,

$$T_1(n, n/(k+1), 6) = \mathcal{O}^*\left(\left(2^{3n/6} + 2^{(1-h(\frac{1}{(k+1)}))n}\right)\right)$$

and

$$|\mathcal{C}| = \mathcal{O}^*(2^{(1-h(\frac{1}{(k+1)}))n}).$$

By Lemma 8.4,

$$T_2(n, n/(k+1)) = k^{\frac{n}{k+1}} \cdot n^{\mathcal{O}(1)}.$$

We choose $\ell = 6$ to make $T_1(n, \alpha n, \ell)$ smaller than $|\mathcal{C}| \cdot T_2$. We select $\alpha = \frac{1}{k+1}$ because this choice minimizes $|\mathcal{C}| \cdot T_2(n, \alpha n)$.

Thus the total running time of the algorithm is

$$T_1\left(n, \frac{n}{k+1}, 6\right) + |\mathcal{C}| \cdot T_2\left(n, \frac{n}{k+1}\right)$$

$$= \mathcal{O}^*\left(\left(2^{3n/6} + 2^{(1-h(\frac{1}{(k+1)}))n} + 2^{(1-h(\frac{1}{(k+1)}))n} \cdot k^{\frac{n}{k+1}}\right)\right)$$

$$= \mathcal{O}^*\left(\left(2^{(1+\frac{1}{k+1}\log_2 \frac{1}{k+1} + \frac{k}{k+1}\log_2 \frac{k}{k+1} + \frac{1}{k+1}\log_2 k)n}\right)\right)$$

$$= \mathcal{O}^*\left(\left(2^{(1-\frac{1}{k+1}\log_2(k+1) + \frac{k}{k+1}\log_2 k - \frac{k}{k+1}\log_2(k+1) + \frac{1}{k+1}\log_2 k)n}\right)\right)$$

$$= \mathcal{O}^*\left(\left(2^{(1+\log_2\frac{k}{k+1})n}\right)\right)$$

$$= \mathcal{O}^*\left(\left(2 - \frac{2}{k+1}\right)^n\right).$$

$\square$

Notes

Theorem 8.3 is due to Schöning [199, 200]. This approach is based on a randomized polynomial time algorithm for 2-SAT of Papadimitriou [169]. Our exposition of the algorithm follows the book of Mitzenmacher and Upfal [157]. For an introductory book on Probability Theory, we refer to the book of Feller [76].

A different randomized approach to solving k-SAT is given by Paturi, Pudlák, Saks and Zane [172]. They obtain an algorithm solving k-SAT with error probability $o(1)$ in time

$$\mathcal{O}^*(2^{n(1-\frac{\mu_k}{k-1}+o(1))}),$$

where μ_k is an increasing sequence with $\mu_3 \approx 1.227$ and $\lim_{k \to \infty} \mu_k = \pi^2/6$. See also [173].

For $k = 3$, the randomized Schöning's algorithm runs in time $\mathcal{O}^*(1.334^n)$. It is possible to improve this bound by combining the algorithms of Schöning and Paturi, Pudlák, Saks and Zane [124], see also [116]. Rolf reduced the running time to $\mathcal{O}^*(1.323^n)$ [186]. The surveys of Schöning [201] and Iwama [122] contain more information on probabilistic algorithms for SAT.

Theorem 8.9 is due to Dantsin, Goerdt, Hirsch, Kannan, Kleinberg, Papadimitriou, Raghavan and Schöning [59]. Since the algorithm of Lemma 8.8 requires exponential space, the algorithm of Theorem 8.9 also needs exponential space. This can be fixed (at the price of more running time) if in the proof of Lemma 8.8 we choose ℓ of size εn, for some $\varepsilon > 0$. Then for every $\delta > 0$ the algorithm from Theorem 8.9 can be modified to run in polynomial space and time

$$\mathcal{O}^*\left(\left(2 - \frac{2}{k+1} + \delta\right)^n\right).$$

The proof that an approximation for MINIMUM SET COVER can be obtained by a greedy algorithm can be found in [114].

For 3-SAT the deterministic algorithm with running time $\mathcal{O}^*(1.5^n)$ was sped up to $\mathcal{O}^*(1.473^n)$ [41] and to $\mathcal{O}^*(1.465^n)$ [196] by improving the running time of Lemma 8.4. Dantsin and Hirsch provide a nice overview of known techniques in [60]. Feder and Motwani obtained randomized algorithms for Constraint Satisfaction Problem (CSP) [74].

Chapter 9
Split and List

In this chapter we discuss several algorithms based on the following approach. There is a number of efficient algorithms for many problems in P. To apply these algorithms on hard problems, we (exponentially) enlarge the size of a hard problem and apply fast polynomial time algorithm on an input of exponential size. The common way to enlarge the problem is to split the input into parts, and for each part to enumerate (or list) all possible solutions to subproblems corresponding to the part. Then we combine solutions of subproblems to solutions of the input of the original problem by making use of a fast polynomial time algorithm.

9.1 Sort and Search

First let us recall that on a vector space $\mathbb{Q}^m$ over the rational numbers $\mathbb{Q}$ one can define a *lexicographical order* denoted by $\prec$. For vectors $\mathbf{x} = (x_1, x_2, \dots, x_m)$, $\mathbf{y} = (y_1, y_2, \dots, y_m) \in \mathbb{Q}^m$, we define that $\mathbf{x} \prec \mathbf{y}$ if and only if there is a $t \in \{1, 2, \dots, m\}$ such that $x_i = y_i$ for all $i < t$ and $x_t < y_t$. For example, $(2,4,8,3) \prec (2,7,2,4)$. We also write $\mathbf{x} \preceq \mathbf{y}$ if $\mathbf{x} \prec \mathbf{y}$ or $\mathbf{x} = \mathbf{y}$.

Before proceeding with the Split & List technique, let us play a bit with the following "toy" problem. In the 2-TABLE problem, we are given 2 tables T_1 and T_2 each being an array of size $m \times k$, and a vector $\mathbf{s} \in \mathbb{Q}^m$. Each table consists of k vectors of $\mathbb{Q}^m$ in such a way that each vector is a column of the array. The question is, if the table contains an entry from the first column and an entry from the second column such that the sum of these two vectors is $\mathbf{s}$?

$$\begin{array}{|ccc|ccc|}
0\ 1\ 4\ 3 & 0\ 1\ 3\ 0 \\
1\ 3\ 4\ 3 & 2\ 1\ 6\ 0 \\
1\ 5\ 4\ 3 & 3\ 1\ 3\ 0
\end{array}$$

Fig. 9.1 An instance of the 2-TABLE problem with entries from $\mathbb{Q}^3$

F.V. Fomin, D. Kratsch, *Exact Exponential Algorithms*, Texts in Theoretical Computer Science. An EATCS Series, DOI 10.1007/978-3-642-16533-7_9,
© Springer-Verlag Berlin Heidelberg 2010

An example of an instance of a 2-TABLE problem is given in Fig. 9.1. For vector $s = \begin{pmatrix} 4 \\ 4 \\ 4 \end{pmatrix}$, there are two solutions ($\begin{pmatrix} 3 \\ 3 \\ 3 \end{pmatrix}$, $\begin{pmatrix} 1 \\ 1 \\ 1 \end{pmatrix}$) and ($\begin{pmatrix} 4 \\ 4 \\ 4 \end{pmatrix}$, $\begin{pmatrix} 0 \\ 0 \\ 0 \end{pmatrix}$).

A trivial solution to the 2-TABLE problem would be to try all possible pairs of vectors. Each comparison takes $\mathcal{O}(m)$, and the number of pairs of vectors is $\mathcal{O}(k^2)$, which would result in running time $\mathcal{O}(mk^2)$. There is a smarter way to solve this problem.

Lemma 9.1. *The* 2-TABLE *problem for tables* T_1 *and* T_2 *of size* $m \times k$ *with entries from* $\mathbb{Q}^m$ *can be solved in time* $\mathcal{O}(mk \log k)$.

Proof. The vectors of the first table are sorted increasingly in lexicographic order and the vectors of the second table are sorted decreasingly in lexicographic order. Two vectors can be compared in time $\mathcal{O}(m)$. Consequently the sorting can be done in time $\mathcal{O}(mk \log k)$.

Now given sorted vector sequences $\mathbf{a}_1 \preceq \mathbf{a}_2 \preceq \cdots \preceq \mathbf{a}_k$ and $\mathbf{b}_k \preceq \mathbf{b}_{k-1} \preceq \cdots \preceq \mathbf{b}_1$, the algorithm finds out whether there are vectors $\mathbf{a}_i$ and $\mathbf{b}_j$ such that $\mathbf{a}_i + \mathbf{b}_j = \mathbf{s}$. More precisely algorithm 2-table, described in Fig. 9.2, outputs all such pairs and its correctness is based on the following observation. If $\mathbf{a}_i + \mathbf{b}_j \prec \mathbf{c}$, then for every $l \geq i$, $\mathbf{a}_i + \mathbf{b}_l \prec \mathbf{c}$, and thus all vectors $\mathbf{b}_l$, $l \geq i$, can be eliminated from consideration. Similarly, if $\mathbf{c} \prec \mathbf{a}_i + \mathbf{b}_j$, then all vectors $\mathbf{a}_l$, $l \geq i$, are eliminated from consideration. The algorithm takes $\mathcal{O}(k)$ steps and the total running time, including sorting, is $\mathcal{O}(mk \log k)$.   □

Algorithm 2-table.
Input: Tables T_1 and T_2 of size $m \times k$ with columns/vectors $\mathbf{a}_i$ in T_1 and $\mathbf{b}_j$ in T_2, and vector $\mathbf{c}$.
Output: All pairs $(\mathbf{a}_i, \mathbf{b}_j)$ such that $\mathbf{a}_i + \mathbf{b}_j = \mathbf{c}$ and $\mathbf{a}_i \in T_1$, $\mathbf{b}_j \in T_2$.

$i := 1; j := 1$
while $i \leq k$ *and* $j \leq k$ **do**
    if $\mathbf{a}_i + \mathbf{b}_j = \mathbf{c}$ **then**
        ⌊ **return** $(\mathbf{a}_i, \mathbf{b}_j)$
    if $\mathbf{a}_i + \mathbf{b}_j \prec \mathbf{c}$ **then**
        ⌊ $i := i + 1$
    if $\mathbf{c} \prec \mathbf{a}_i + \mathbf{b}_j$ **then**
        ⌊ $j := j + 1$

Fig. 9.2 Algorithm 2-table

Let us remark that with a simple modification that outputs all pairs $\mathbf{a}_i + \mathbf{b}_j = \mathbf{c}$ and increments counters i and j, the algorithm can enumerate all solutions $(\mathbf{a}_i, \mathbf{b}_j)$ within the same running time.

The solution of surprisingly many hard problems can be reduced to the solution of the 2-TABLE problem. The main idea of the approach is to partition an input of a problem into two subproblems, solve them separately and find the solution to the original problem by combining the solutions to the subproblems. We consider three NP-hard problems.

Subset Sum. In the SUBSET SUM problem, we are given positive integers $a_1, a_2, \ldots, a_n$, and S. The task is to find a subset $I \subseteq \{1, 2, \ldots n\}$ such that

$$\sum_{i \in I} a_i = S,$$

or to report that no such subset exists. For example, for $a_1 = 5$, $a_2 = 5$, $a_3 = 10$, $a_4 = 60$, $a_5 = 61$, and $S = 70$, the solution is $I = \{1, 2, 4\}$.

Theorem 9.2. *The* SUBSET SUM *problem can be solved in time* $\mathcal{O}(n2^{n/2})$.

Proof. We partition $\{a_1, a_2, \ldots, a_n\}$ into two sets $X = \{a_1, a_2, \ldots, a_{\lfloor n/2 \rfloor}\}$ and $Y = \{a_{\lfloor n/2 \rfloor+1}, \ldots, a_n\}$. For each of these two sets, we compute the set of all possible subset sums. The total number of computed sums is at most $2^{n/2+1}$. Let I_X and I_Y be the sets of computed sums for X and Y respectively (let us remark that 0 belongs to both I_X and I_Y). Then there is a solution to the SUBSET SUM problem if and only if there is an $s_X \in I_X$ and an $s_Y \in I_Y$ such that $s_X + s_Y = S$. To find such s_X and s_Y, we reduce the problem to an instance of the 2-TABLE problem. We build an instance of the 2-TABLE. Table T_1 is formed by the elements of I_X and table T_2 is formed by the elements of I_Y. Both are arrays of size $m \times k$, where $m = 1$ and $k \leq 2^{n/2}$. Then by Lemma 9.1, we can find two elements, one from each table, whose sum is S (if they exist) in time $\mathcal{O}(2^{n/2} \log 2^{n/2}) = \mathcal{O}(n2^{n/2})$. $\qquad \square$

Exact Satisfiability. In the EXACT SATISFIABILITY problem (XSAT), we are given a CNF-formula F with n variables and m clauses. The task is to find a satisfying assignment of F such that each clause contains exactly one true literal. For example, the CNF formula

$$(x_1 \vee x_2 \vee \overline{x_3}) \wedge (\overline{x_1} \vee \overline{x_2}) \wedge (\overline{x_1} \vee x_3)$$

is satisfied by the truth assignment $x_1 = $ true, $x_2 = $ false, and $x_3 = $ true, moreover, for this assignment, each clause is satisfied by exactly one literal.

While there are faster branching algorithms for XSAT, we find this example interesting because its comparison with SAT helps us to better understand which kind of properties are necessary to reduce a problem to the 2-TABLE problem.

Theorem 9.3. *The problem* XSAT *is solvable in time* $\mathcal{O}^*(2^{n/2})$.

Proof. Let F be an input of XSAT. Let its set of clauses be $\{c_1, c_2, \ldots, c_m\}$ and let its set of variables be $\{x_1, x_2, \ldots, x_n\}$. We split the variables into two sets $X = \{x_1, x_2, \ldots, x_{\lfloor n/2 \rfloor}\}$ and $Y = \{x_{\lfloor n/2 \rfloor+1}, \ldots, x_n\}$. For every possible truth assignment f of the variables of X which assigns to each variable either the value true or false, we form its characteristic vector $\chi(f, X) \in \mathbb{Q}^m$. The ith coordinate of $\chi(f, X)$ is equal to the number of literals which evaluate to true in the clause c_i. Similarly, for every possible truth assignment g of the variables of Y we form its characteristic vector $\chi(g,Y) \in \mathbb{Q}^m$. The jth coordinate of $\chi(g,Y)$ is equal to the number of literals which evaluate to true in the clause c_j.

Let us note that the input formula F is exactly satisfied if and only if there is an assignment f of X and an assignment g of Y such that $\chi(f,X) + \chi(g,Y) = (1,1,\ldots,1)$. We form two tables: table T_1 contains characteristic vectors of X and table T_2 contains characteristic vectors of Y. Each table has at most $2^{\lceil n/2 \rceil}$ columns. Thus we can again apply Lemma 9.1, and solve XSAT in time $\mathcal{O}^*(2^{n/2})$. $\qquad\square$

Why can this approach not be used to solve SAT in time $\mathcal{O}^*(2^{n/2})$? This is because by constructing an instance of the 2-TABLE for an instance of SAT the same way as we did for XSAT, we have to find characteristic vectors such that $(1,1,\ldots,1) \preceq \chi(f,X) + \chi(g,Y)$. This is a real obstacle, because we cannot use Lemma 9.1 anymore: the argument "if $\mathbf{a}_i + \mathbf{b}_j \prec (1,1,\ldots,1)$, then for every $l \geq i$, $\mathbf{a}_i + \mathbf{b}_l \prec (1,1,\ldots,1)$" does not hold anymore. In the worst case (without having any ingenious idea) we have to try all possible pairs of vectors.

Knapsack. In the BINARY KNAPSACK problem, we are given a positive integer W and n items $s_1, s_2, \ldots, s_n$, each item has its value a_i and its weight w_i, which are positive integers. The task is to find a subset of items of maximum total value subject to the constraint that the total weight of these items is at most W.

To solve the BINARY KNAPSACK problem in time $\mathcal{O}^*(2^{n/2})$, we reduce its solution to the solution of the following MODIFIED 2-TABLE problem. We are given two tables T_1 and T_2 each one an array of size $1 \times k$ whose entries are positive integers, and an integer W. The task is to find one number from the first and one from the second table whose sum is at most W. An example is given in Fig. 9.3.

$$\boxed{10\ 2\ 4\ 12}\ \boxed{15\ 6\ 11\ 14}$$

Fig. 9.3 In this example, for $W = 14$, the solution is the pair of integers $(2,11)$.

The problem MODIFIED 2-TABLE can be solved in time $\mathcal{O}(k \log k)$ with an algorithm similar to the one for 2-table.

In the first step, the algorithms sorts the entries of T_1 in increasing order and the ones of T_2 in decreasing order. Let x be an entry in T_1 and y an entry in T_2. We observe the following. If $x + y \leq W$, then for all z appearing after y in T_2, we have $z < y$, and, consequently, $x + z \leq x + y$. Therefore, all such pairs (x,z) can be eliminated from consideration, as they cannot provide a better answer than (x,y). Similarly, if $x + y > W$ then for all z appearing after x in T_1, $z + y > W$, and thus all pairs (z,y), $z \geq x$, can be eliminated from consideration. Thus after sorting, which requires $\mathcal{O}(k \log k)$ time, one can find the required pair of numbers in $\mathcal{O}(k)$ steps. This observation is used to prove the following theorem.

Theorem 9.4. *The* BINARY KNAPSACK *problem is solvable in time* $\mathcal{O}^*(2^{n/2})$.

Proof. To solve the BINARY KNAPSACK problem, we split the set of items into two subsets $s_1, s_2, \ldots, s_{\lfloor n/2 \rfloor}$ and $s_{\lfloor n/2 \rfloor + 1}, \ldots, s_n$, and for each subset $I \subseteq \{1, 2, \ldots, \lfloor n/2 \rfloor\}$, we construct a couple $x_I = (A_I, W_I)$, where

$$A_I = \sum_{i \in I} a_i, \text{ and } W_I = \sum_{i \in I} w_i.$$

Thus we obtain a set X of couples and the cardinality of X is at most $2^{n/2}$. Similarly we construct the set Y which consists of all couples $y_J = (A_J, W_J)$, where $J \subseteq \{\lfloor n/2 \rfloor + 1, \lfloor n/2 \rfloor + 1 + 1, \ldots, n\}$. Then the problem boils down to finding couples $x_I \in X$ and $y_J \in Y$ such that $A_I + A_J$ is maximum subject to the constraint $W_I + W_J \leq W$.

To reduce the problem to an instance of the MODIFIED 2-TABLE problem discussed above, we perform the following preprocessing: a couple (A_I, W_I) is removed from X (or Y) if there is a couple $(A_{I'}, W_{I'})$, $I \neq I'$, from the same set such that $A_{I'} \geq A_I$ and $W_{I'} \leq W_I$. The argument here is that the set of items with couple $(A_{I'}, W_{I'})$ has higher value and smaller weight, so we prefer $(A_{I'}, W_{I'})$ and can safely remove (A_I, W_I) from X. In the case of $(A_I, W_I) = (A_{I'}, W_{I'})$, we break ties arbitrarily. In other words, we remove couples dominated by some other couple.

This preprocessing is done in time $\mathcal{O}^*(n 2^{n/2})$ in the following way for X and similarly for Y. First the items of the set X (or Y) are sorted in increasing order according to their weights. At the second step of the preprocessing we are given a list of couples sorted by increasing weights

$$(A_1, W_1), (A_2, W_2), \cdots, (A_k, W_k),$$

where $k \leq 2^{n/2}$ and for every $1 \leq i < j \leq k$, $W_i \leq W_j$. We put $A := A_1$ and move in the list from 1 to k performing the following operations: if $A_i > A$, we put $A := A_i$. Otherwise ($A_i \leq A$), we remove (A_i, W_i) from the list. This procedure takes $\mathcal{O}(k)$ steps and as the result of it we have produced a set of couples with no couple dominated by any other one.

Thus after preprocessing done for X and for Y, we have that $(A_I, W_I) \in X$ and $(A_J, W_J) \in Y$ have maximum sum $A_I + A_J$ subject to $W_I + W_J \leq W$ if and only if the sum $W_I + W_J$ is maximum subject to $W_I + W_J \leq W$. What remains is to construct the table of size $2 \times 2^{n/2}$ and use the algorithm for the MODIFIED 2-TABLE problem. This step requires time $\mathcal{O}^*(2^{n/2})$. This concludes the proof. □

A natural idea to improve the running time of all algorithms based on reductions to the k-TABLE problem, is to partition the original set into $k \geq 3$ subsets and reduce to the k-TABLE problem. However, it is not clear how to use this approach to obtain better overall running times. Consider the following k-TABLE problem: given k tables $T_1, T_2, \ldots T_k$ such that each table is an array of size $m \times k$ with entries from $\mathbb{R}^m$, the task is for a given vector $\mathbf{c}$, to find a set of vectors $(\mathbf{c}_1, \mathbf{c}_2, \ldots, \mathbf{c}_k)$, $\mathbf{c}_i \in T_i$, such that

$$\mathbf{c}_1 + \mathbf{c}_2 + \cdots + \mathbf{c}_k = \mathbf{c}.$$

We can solve the k-TABLE problem in time $\mathcal{O}(n^{k-1} + kn\log n)$ by recursively applying the algorithm for the 2-TABLE problem. Unfortunately we do not know any faster algorithm for this problem.

Thus if we split an instance of a hard problem, like XSAT, into k subsets, construct $2^{n/k}$ sets of vectors for each table, and use an algorithm for solving the k-TABLE problem, we obtain an algorithm of running time $\mathcal{O}^*(2^{(k-1)n/k})$.

However, the idea of reducing to a k-TABLE problem can be useful to reduce the space required by such algorithms. All algorithms discussed in this section keep tables of sizes $m \times 2^{n/2}$ and thus the space needed is $2^{n/2}$. Schroeppel and Shamir [202] used the k-TABLE problem to reduce the space requirement of such algorithms to $2^{n/4}$.

9.2 Maximum Cut

In this section we describe an algorithm due to Williams solving the MAXIMUM CUT problem. The algorithm is based on a fast way of finding triangles in a graph. This approach is based on fast square matrix multiplication. Let us recall, that the product of two $n \times n$ matrices can be computed in $\mathcal{O}(n^{\omega})$ time, where $\omega < 2.376$ is the so-called square matrix multiplication exponent.

Maximum Cut. In the MAXIMUM CUT problem (Max-Cut), we are given an undirected graph $G = (V,E)$. The task is to find a set $X \subseteq V$ maximizing the value of $CUT(X, V \setminus X)$, i.e. the number of edges with one endpoint in X and one endpoint in $V \setminus X$.

While a naive way of finding a triangle in a graph would be to try all possible triples of vertices, there is a faster algorithm for doing this job.

Theorem 9.5. *A triangle in a graph on n vertices can be found in time $\mathcal{O}(n^{\omega})$ and in $\mathcal{O}(n^2)$ space.*

Proof. Let $A(G)$ be the adjacency matrix of G. It is easy to prove that in the kth power $(A(G))^k$ of $A(G)$ the entry $(A(G))^k[i,i]$ on the main diagonal of $(A(G))^k$ is equal to the number of walks of length k which start and end in vertex i. Every walk of length 3 which starts and ends at i must pass through 3 vertices, and thus is a triangle. We conclude that G contains a triangle if and only if $(A(G))^3$ has a non-zero entry on its main diagonal. The space required to compute the product of matrices is proportional to the size of $A(G)$, which is n^2. $\square$

Theorem 9.6. *The MAXIMUM CUT problem on n-vertex graphs is solvable in time $\mathcal{O}^*(2^{\omega n/3}) = \mathcal{O}(1.7315^n)$, where $\omega < 2.376$ is the square matrix multiplication exponent.*

Proof. Let us assume that $G = (V,E)$ is a graph on n vertices and that n is divisible by 3. (If not we can add one or two isolated vertices which do not change the value of

the maximum cut and add a polynomial factor to the running time of the algorithm.)
Let V_0, V_1, V_2 be an arbitrary partition of V into sets of sizes $n/3$.

We construct an auxiliary weighted directed graph $A(G)$ as follows. For every
subset $X \subseteq V_i$, $0 \le i \le 2$, the graph $A(G)$ has a vertex X. Thus $A(G)$ has $3 \cdot 2^{n/3}$
vertices. The arcs of $A(G)$ are all possible pairs of the form (X, Y), where $X \subseteq V_i$,
$Y \subseteq V_j$, and $j = i + 1 \pmod 3$. Thus $A(G)$ has $3 \cdot 2^{2n/3}$ arcs. For every arc (X, Y)
with $X \subseteq V_i$ and $Y \subseteq V_j$, $i \ne j$, we define its weight

$$w(X, Y) = CUT(X, V_i \setminus X) + CUT(X, V_j \setminus Y) + CUT(Y, V_i \setminus X).$$

Claim. The following properties are equivalent

(i)   There is $X \subseteq V$ such that $CUT(X, V \setminus X) = t$.
(ii) The auxiliary graph $A(G)$ contains a directed triangle X_0, X_1, X_2, $X_i \subseteq V_i$, $0 \le i \le 2$, such that

$$t = w(X_0, X_1) + w(X_1, X_2) + w(X_2, X_0).$$

Proof (Proof of Claim). To prove $(i) \Rightarrow (ii)$, we put $X_i = X \cap V_i$, $0 \le i \le 2$. Then
every edge e of G contributes 1 to the sum $w(X_0, X_1) + w(X_1, X_2) + w(X_2, X_0)$ if
e is an edge between X and $V \setminus X$, and 0 otherwise. To prove $(ii) \Rightarrow (i)$, we put
$X = X_0 \cup X_1 \cup X_2$. Then again, every edge is counted in $CUT(X, V \setminus X)$ as many
times as it is counted in $w(X_0, X_1) + w(X_1, X_2) + w(X_2, X_0)$. $\square$

To find out whether the condition (ii) of the claim holds, we do the following.
We try all possible values of $w(X_i, X_j)$, $j = i + 1 \pmod 3$. Thus for every triple $W = (w_{01}, w_{12}, w_{20})$ such that $w = w_{01} + w_{12} + w_{20}$, we consider the subgraph $A(G, W)$
of $A(G)$ which contains only the arcs of weight w_{ij} from $X_i \subseteq V_i$ to $X_j \subseteq V_j$. For
every value of t, the number of such triples is at most t^3. The subgraph $A(G, W)$ can
be constructed in time $\mathcal{O}^*(2^{2n/3})$ by going through all arcs of $A(G)$. But then there
exists a triple W satisfying (ii) if and only if the underlying undirected graph of
$A(G, W)$ contains a triangle of weight W. By Theorem 9.5, verifying whether such
a triangle exists can be done in time $\mathcal{O}^*(2^{\omega n/3})$. Thus for every value of t, we try all
possible partitions of t, and for each such partition we construct the graph $A(G, W)$
and check whether it contains a triangle of weight t. The total running time is

$$\mathcal{O}^*(t \cdot t^3 (2^{\omega n/3} + 2^{2n/3})) = \mathcal{O}^*(2^{\omega n/3}).$$

Notes

The name Split and List for the technique is due to Ryan Williams, who used it in his
PhD thesis [217]. The algorithms for SUBSET SUM and BINARY KNAPSACK are
due to Horowitz and Sahni [117] and Schroeppel and Shamir [202]. Note that this is
an early paper of Adi Shamir, one of the three inventors of the RSA public-key cryp-
tosystem. The space requirements in these algorithms can be improved to $\mathcal{O}^*(2^{n/4})$
while keeping the same running time of $\mathcal{O}^*(2^{n/2})$ [202]. Howgrave-Graham and

Joux improve the algorithm of Schroeppel and Shamir for SUBSET SUM on random inputs i [118].

Fomin, Golovach, Kratochvil, Kratsch and Liedloff used the Split and List approach to list different types of (σ, ρ) dominating sets in graphs [82]. Klinz and Woeginger used this approach for computing power indices in voting games [131].

Williams [216] provides a variant of Theorem 9.6 for solving a more general counting version of WEIGHTED 2-CSP (a variant of constraint satisfaction with constraints of size at most 2). Williams' PhD thesis [217] contains further generalizations of this approach.

Theorem 9.5 is due to Itai and Rodeh [121]. A natural question concerning the proof of Theorem 9.6 is, whether partitioning into more than three parts would be useful. The real obstacle is the time spent to find a clique of size k in a graph. Despite many attempts, the following result of Nešetřil & Poljak was not improved for more than 25 years: The number of cliques of size $3k$ in an n-vertex graph can be found in time $\mathcal{O}(n^{\omega k})$ and space $\mathcal{O}(n^{2k})$ [163]. Eisenbrand and Grandoni [67] succeeded in improving the result of Nešetřil and Poljak for a $(3k+1)$-clique and a $(3k+2)$-clique for small values of k. In particular, they show how to find a cliques of size 4,5, and 7 in time $\mathcal{O}(n^{3.334})$ $\mathcal{O}(n^{4.220})$, and $\mathcal{O}(n^{5.714})$, respectively.

The first algorithm that performs a matrix multiplication faster than the standard Gaussian elimination procedure implying $\omega \leq \log_2 7 < 2.81$ is due to Strassen [210]. The proof that $\omega < 2.376$ is due to Coppersmith and Winograd [51].

Chapter 10
Time Versus Space

We have already met different types of exponential algorithms. Some of them use only polynomial space, among them in particular the branching algorithms. On the other hand, there are exponential time algorithms needing exponential space, among them in particular the dynamic programming algorithms. In real life applications polynomial space is definitely preferable to exponential space. However, often a "moderate" usage of exponential space can be tolerated if it can be used to speed up the running time. Is it possible by sacrificing a bit of running time to gain in space? In the first section of this chapter we discuss such an interpolation between the two extremes of space complexity for dynamic programming algorithms. In the second section we discuss an opposite technique to gain time by using more space, in particular for branching algorithms.

10.1 Space for Time: Divide & Conquer

We present a technique to transform an exponential space algorithm into a polynomial space algorithm. It is a Divide & Conquer approach that leads to recursive algorithms. Typically the technique is applied to dynamic programming algorithms needing exponential space. It can be seen as based on the central idea of Savitch's proof that non-deterministic polynomial space is equal to deterministic polynomial space.

We demonstrate this approach for the TRAVELLING SALESMAN problem (TSP). In Chap. 1, we gave a dynamic programming algorithm solving TSP on n cities in time and space $\mathcal{O}^*(2^n)$. It is possible to solve TSP by applying ideas based on inclusion-exclusion, in almost the same way as we did for counting Hamiltonian paths in Chap. 4. However, this approach yields an algorithm that runs in time $\mathcal{O}^*(W2^n)$ and uses space $\mathcal{O}^*(W)$, where W is the maximum distance between two cities. Since W can be exponential in n, this does not give a polynomial space algorithm.

F.V. Fomin, D. Kratsch, *Exact Exponential Algorithms*, Texts in Theoretical
Computer Science. An EATCS Series, DOI 10.1007/978-3-642-16533-7_10,
© Springer-Verlag Berlin Heidelberg 2010

Theorem 10.1. *The* TRAVELLING SALESMAN *problem on n cities is solvable in time* $\mathcal{O}^*(4^n n^{\log n})$ *using only polynomial space.*

Proof. Let us consider an instance of TSP with a set of distinct cities

$$C = \{c_1, c_2, \ldots, c_n\}.$$

Let $d(c_i, c_j)$ be the distance between cities c_i and c_j. The algorithm is based on the following simple idea. We try to guess the first half of the cities visited by the salesman. The number of all such guesses is at most 2^n and for each of the guesses we solve two problems of size $n/2$ recursively.

For a nonempty subset S of C and cities $c_i, c_j \in S$, we define $OPT[S, c_i, c_j]$ to be the minimum length of a tour which starts in c_i, visits all cities of S and ends in c_j. When $|S| = 1$, we put (slightly abusing notations) $OPT[S, c_i, c_j] = 0$. For $S = \{c_i, c_j\}$, $i \neq j$, we put

$$OPT[S, c_i, c_j] = d(c_i, c_j).$$

For a subset $S \subseteq C$, of size at least 3, $c_i, c_j \in S$, let P be a tour of length $OPT[S, c_i, c_j]$ which starts in c_i, visits all cities in S and ends in c_j. Let S_1 be the first half of the cities visited by P, x be the last city in S_1 visited by P and y be the first city not in S_1 visited by P. Then the length of the part of P from c_i to x is $OPT[S_1, c_i, x]$, the length of the part from y to c_j is $OPT[S \setminus S_1, y, c_j]$, and the total length of P is

$$OPT[S_1, c_i, x] + OPT[S \setminus S_1, y, c_j] + d(x, y).$$

Therefore, for every subset $S \subseteq C$ of size at least 3 and every pair $c_i, c_j \in S$, $i \neq j$, we can compute the value of $OPT[S, c_i, c_j]$ by making use of the following formula.

$$OPT[S, c_i, c_j] = \min_{\substack{S' \subset S \setminus \{c_j\} \\ c_i \in S' \\ |S'| = \lceil |S|/2 \rceil \\ x \in S', y \in S \setminus S'}} OPT[S', c_i, x] + OPT[S \setminus S', y, c_j]$$

$$+ d(x, y). \qquad (10.1)$$

Then the length of an optimal tour is equal to

$$\min_{c_i, c_j \in C, i \neq j} OPT[C, c_i, c_j] + d(c_i, c_j). \qquad (10.2)$$

Now to compute the optimal tour, we try all possible pairs of cities and use (10.1) recursively to compute the values in (10.2).

Let $T(n)$ be the maximum number of calls in the recursive algorithm based on (10.1). For a set S of size k and fixed cites c_i, c_j, we try all possible subsets S' of S of size $\lceil k/2 \rceil$ containing c_i and not containing c_j. The number of such sets is at most 2^k. For every such subset S' there are at most $\lceil k/2 \rceil$ values of $OPT(S', c_i, x)$ and at most $\lfloor k/2 \rfloor$ values of $OPT(S \setminus S', y, c_j)$. Thus we can estimate $T(n)$ by the following recursive formula

$$T(n) \leq 2^n \left(\lceil n/2 \rceil \cdot T(\lceil n/2 \rceil) + \lfloor n/2 \rfloor \cdot T(\lfloor n/2 \rfloor) \right)$$
$$= \mathcal{O}(2^{n(1+1/2+1/4+1/8+\cdots)} n^{\log n} T(1))$$
$$= \mathcal{O}(4^n n^{\log n}).$$

The recursion requires a polynomial number of steps for every recursive call and thus the total running time of the algorithm is $\mathcal{O}^*(4^n n^{\log n})$.

The space required on each recursion level to enumerate all sets S' is polynomial. Since the recursion depth is $\mathcal{O}(\log n)$, the polynomial space bound is easily met. □

Let us remark that using a log-cost RAM model for the analysis in the proof of Theorem 10.1 leads to a running time of $\mathcal{O}(4^n n^{\log n} \log W)$ and space $(n \log W)^{\mathcal{O}(1)}$.

The two examples of algorithms for solving the TRAVELLING SALESMAN problem either by Dynamic Programming or by Divide & Conquer are two extremes of time-space exchange. The dynamic programming algorithm needs $\mathcal{O}^*(2^n)$ time and space, and for this algorithm the product of space and time is

$$SPACE \times TIME = \mathcal{O}^*(4^n).$$

The Divide & Conquer algorithm has almost the same product of space and time, namely $4^{n+o(n)}$.

By combining these two approaches it is possible to balance the exchange of time and space.

Theorem 10.2. *For every* $i \in \{0, 1, 2, \ldots, \lceil \log_2 n \rceil\}$, *the problem* TSP *on n cities is solvable in time* $\mathcal{O}^*(2^{n(2-1/2^i)} n^i)$ *and space* $\mathcal{O}^*(2^{n/2^i})$.

Proof. The proof of the theorem is almost the same as the proof of Theorem 10.1. It is also based on (10.1) and the only difference is that we stop recursion when the size of the problem is $\lceil n/2^i \rceil$. For sets S of size $\lceil n/2^i \rceil$ the dynamic programming algorithm computes $OPT[S, c_i, x]$ in time and space $\mathcal{O}^*(2^{|S|}) = \mathcal{O}^*(2^{n/2^i})$. There are i levels of recursion and the number of recursive calls is

$$T(n) \leq \mathcal{O}(2^{n(1+1/2+1/4+1/8+\cdots+1/2^{i-1})} n^i)$$
$$= \mathcal{O}(2^{2n(1-1/2^i)} n^i).$$

For each recursive call we use time and space $\mathcal{O}^*(2^{n/2^i})$, and the running time of the algorithm is

$$\mathcal{O}^*(2^{n/2^i} \cdot 2^{2n(1-1/2^i)} n^i) = \mathcal{O}^*(2^{n(2-1/2^i)} n^i).$$

□

Let us remark that the case $i = 0$ corresponds to the running time and space requirements of the dynamic programming algorithm of Chap. 1 and the case $i = \lceil \log_2 n \rceil$ corresponds to Theorem 10.1.

The product of space and time required by algorithms in Theorem 10.2 is $4^{n+o(n)}$. The following result of Koivisto and Parviainen gives an algorithm for TSP, such that the product of space and time required is $\mathcal{O}^*(3.9271^n)$.

Theorem 10.3. *TSP on n cities is solvable in time $\mathcal{O}^*(2.7039^n)$ and space $\mathcal{O}^*(1.4525^n)$.*

Proof. Let us assume for simplicity that n is divisible by 26. We divide n cities into $n/26$ sets $S_1, S_2, \ldots, S_{n/26}$ of equal size. For each set S_i we consider all possible partitions (A_i, B_i) into sets of equal size, i.e. of size 13. Thus for each set we have $\binom{26}{13}$ partitions and in total we consider $\binom{26}{13}^{n/26}$ partitions. For each such partition, we consider orderings of the cities in which for every $i \in \{1, 2, \ldots, n/26\}$, all cities from A_i are before any city from B_i. Since we try all possible partitions, we are quarantined that at least for one of the partitions at least one ordering with A_i being before B_i is also an ordering corresponding to the optimum tour of the salesman.

For each of the partitions, we perform dynamic programming almost identically to the algorithm of Chap. 1. For a fixed set of partitions

$$(A_1, B_1), \ldots, (A_{n/26}, B_{n/26}),$$

we first compute all values $OPT[S, c_i]$, where $i \leq n/2$ and $S \cup \{c_i\}$ is a subset of cities from $A_1 \cup A_2 \cup \cdots \cup A_{n/26}$. This computation requires time and space $\mathcal{O}^*(2^{n/2})$. We proceed with dynamic programming over all subsets S such that for every $i \in \{1, 2, \ldots, n/26\}$, a subset of B_i can be a part of S only if all vertices of A_i are in S. Thus S can contain

- either one of the $2^{13} - 1$ proper subsets of A_i (but not A_i) and none of the vertices from B_i,
- all vertices of A_i and one of the 2^{13} subsets of B_i.

The running time and space required to compute an optimal solution for each fixed set of partitions is up to a polynomial factor proportional to the amount of such subsets S, which is

$$\mathcal{O}^*((2^{13} - 1 + 2^{13})^{n/26}) = \mathcal{O}^*(1.4525^n).$$

Thus by trying all possible partitions and running the dynamic programming algorithm on each try, we obtain solve TSP in time

$$\mathcal{O}^*(\binom{26}{13}^{n/26} \cdot (2^{13} - 1 + 2^{13})^{n/26}) = \mathcal{O}^*(2.7039^n).$$

and space $\mathcal{O}^*(1.4525^n)$. $\square$

Dynamic programming algorithms of running time $\mathcal{O}^*(2^n)$ for various problems have been mentioned in Chap. 3, among them SCHEDULING, CUTWIDTH, OPTIMAL LINEAR ARRANGEMENT, and TREEWIDTH. For all of these, $\mathcal{O}^*(4^n n^{\log n})$ time polynomial space algorithms can be established by Divide & Conquer. Of course, the result of Theorem 10.3 also holds for a large variety of permutation problems.

Our second example is the COLORING problem. Before showing how to apply the Divide & Conquer technique to a dynamic programming algorithm, we present a helpful combinatorial lemma.

Lemma 10.4. *Let* $N = \{n_1, \ldots, n_k\}$ *be any non-empty multiset of positive integers, let* $n = \sum_{j=1}^k n_j$ *be the sum of the elements in N, and let m be a largest element of N. Then* $N \setminus \{m\}$ *can be partitioned into multisets* C_1, C_2 *such that for* $i = 1, 2$:

$$\sum_{n_j \in C_i} n_j \leq n/2.$$

Proof. Take any order of the elements in the multiset $N \setminus \{m\}$ and renumber the elements $n_1, n_2, \ldots, n_{k-1}$. Then find the unique index r such that $n_1 + \cdots + n_{r-1} \leq n/2$ and $n_1 + \cdots + n_r > n/2$, and thus $n_{r+1} + \cdots n_{k-1} + m \leq n/2$. Combined with $n_r \leq m$ this implies $n_r + n_{r+1} + \cdots n_{k-1} \leq n/2$. Consequently, $C_1 = \{n_1, \ldots, n_r\}$ and $C_2 = \{n_{r+1}, \ldots, n_{k-1}\}$ is the desired partition. $\qquad\square$

A dynamic programming algorithm to compute the chromatic number of a graph and also a minimum coloring has been presented in Chap. 3. This algorithm has running time $\mathcal{O}^*((1 + \sqrt[3]{3})^n) = \mathcal{O}(2.4423^n)$ and needs exponential space. The $\mathcal{O}^*(2^n)$ algorithm from Chap. 4 is based on inclusion-exclusion and also requires exponential space. By Corollary 4.14, there is also a polynomial space algorithm computing the chromatic number of a graph which runs in time $\mathcal{O}^*(2.2461^n)$. The polynomial space algorithm based on Divide & Conquer for graph coloring does not provide such a good running time and is included here only as an example demonstrating the technique.

To apply Divide & Conquer to the dynamic programming coloring algorithm, we observe that any coloring, and in particular a minimum coloring, is a partition of the graph into independent sets, where one independent set might be assumed to be a maximal one. Hence to compute the chromatic number and a minimum coloring we guess the "middle" independent set I that we may assume to be a largest independent set of the coloring and also a maximal independent set of G, and C_1 as the union of the "left" independent sets in the minimum coloring and C_2 as the union of the "right" independent sets. Then the algorithm recursively computes a minimum coloring of $G[C_1]$ and also one of $G[C_2]$. Finally minimum colorings of $G[C_1]$ and $G[C_2]$ combined with I will provide a minimum coloring of G with respect to the partition (C_1, I, C_2).

There is another important feature we have to take care of. The best running time of Divide & Conquer is obtained when we recurse on balanced subproblems. For this we want to work only with partitions (C_1, I, C_2) satisfying $|C_1| \leq n/2$ and $|C_2| \leq n/2$. To guarantee this, Lemma 10.4 is used. Let us assume that the minimum coloring of G is $I_1, I_2, \ldots, I_k$. W.l.o.g. the largest independent set is a maximal one, say $I = I_k$. Now we use Lemma 10.4 with the multiset $N = \{|I_1|, |I_2|, \ldots, |I_k|\}$ and $m = |I_k|$. Then there is an index r such that $C_1 = \{|I_1|, \ldots, |I_{r-1}|\}$, $C_2 = \{|I_r|, \ldots, |I_{k-1}|\}$, $|C_1| \leq n/2$ and $|C_2| \leq n/2$.

Hence the following algorithm correctly computes a minimum coloring of G: Recursively compute a minimum coloring for $G[C_1]$ and $G[C_2]$ for all partitions

(C_1, I, C_2) where I is a maximal independent set of G, $|C_1| \leq n/2$ and $|C_2| \leq n/2$. For each partition (C_1, I, C_2), combining a minimum coloring of $G[C_1]$ and $G[C_2]$ with the maximal independent set I gives a coloring of G. Minimizing over all partitions provides a minimum coloring of G as shown above.

It remains to analyze the running time and the space needed by the algorithm. To analyze the running time let $T(n)$ be the number of recursive calls the algorithm makes on an n-vertex graph. The algorithm has to enumerate all maximal independent sets of G. By a well-known result of Moon and Moser [161] (Theorem 1.1), a graph on n vertices has at most $3^{n/3}$ maximal independent sets and they can be enumerated in time $3^{n/3} n^{O(1)}$ and polynomial space, as mentioned in Chap. 1. The number of sets C_i is at most 2^n. Taking into account the polynomial overhead for computing all partitions (C_1, I, C_2), we establish the recurrence

$$T(n) \leq 3^{n/3} 2^n (T(n/2) + T(n/2)).$$

Note that this recurrence is similar to the one in the algorithm for TSP. It can be solved in the same way and one obtains

$$T(n) = \mathcal{O}^*((3^{1/3} \cdot 2)^{2n}) = \mathcal{O}^*(8.3204^n).$$

Every procedure called recursively can be performed in polynomial time when recursive calls are not considered. Thus the running time of the algorithm is the number of recursive calls times a polynomial factor, and thus $\mathcal{O}^*(8.3204^n)$.

The number of recurrence levels is $\mathcal{O}(\log n)$, and at each step of the recursion we use a polynomial space algorithm to enumerate all maximal independent sets and the subsets C_1 and C_2. Thus the space used by the algorithm is polynomial. We conclude with the following theorem.

Theorem 10.5. *There is a polynomial space algorithm computing a minimum coloring of the input n-vertex graph in time $\mathcal{O}(8.3204^n)$.*

It is possible to establish a result for the COLORING problem similar to Theorem 10.2. For example, one can stop the recursion after reaching problems with instances on $n/2^i$ vertices and apply to such graphs the $2^{n/2^i}$ algorithm from Chap. 4. By exactly the same arguments as for the TSP problem, we obtain the following theorem.

Theorem 10.6. *For every $i \in \{0, 1, 2, \ldots, \lceil \log_2 n \rceil\}$, a minimum coloring of an n-vertex graph can be computed in time $\mathcal{O}^*(3^{\frac{2n(1-1/2^i)}{3}} 2^{n(2-1/2^i)} n^i)$ and space $\mathcal{O}^*(2^{n/2^i})$.*

10.2 Time for Space: Memorization

The time complexity of many exponential time branching algorithms, that originally need only polynomial space, can be reduced at the cost of exponential space complexity via the memorization technique. The method was first applied to an exact algorithm for the MAXIMUM INDEPENDENT SET problem by Robson[185]. Memorization allows us to establish the best exponential time algorithm for various well-studied NP-hard problems e.g. MAXIMUM INDEPENDENT SET.

The basic idea of memorization is the following: In branching algorithms, especially on the lower levels of branching, isomorphic instances of the problem can occur exponentially many times in different branches. To avoid recomputation on the same instance, the solutions of all the subproblems are stored in an (exponential size) database. If the same subproblem turns up more than once, then the algorithm does not solve it for a second time, but looks up the already computed result. The database is implemented in such a way that the query time is logarithmic in the number of solutions stored and thus polynomial in the size of the problem: this way the cost of each look-up is polynomial. The techniques described in this subsection can easily be adapted to many other branching algorithms.

To illustrate how to apply memorization to a branching algorithm we consider a specific NP-hard problem and a specific algorithm to solve it. Morevover, in our first example we analyze it with a simple measure.

Let us first recall the definition of the problem. In the MAXIMUM INDEPENDENT SET problem (MIS) we are given an undirected and simple graph $G = (V, E)$. The task is to construct an independent set of maximum cardinality. Recall that an independent set of G is a vertex set S such that no two vertices of S are adjacent.

We describe how to apply memorization to a branching algorithm using the algorithm of Fig. 10.1.

Algorithm mis5(G).
Input: A graph $G = (V, E)$.
Output: The maximum cardinality of an independent set of G.

if $\Delta(G) \geq 3$ then
    choose a vertex v of maximum degree in G
    return $\max(1 + \text{mis5}(G \setminus N[v]), \text{mis5}(G \setminus v))$
if $\Delta(G) \leq 2$ then
    compute $\alpha(G)$ using a polynomial time algorithm
    return $\alpha(G)$

Fig. 10.1 Algorithm mis5 for MAXIMUM INDEPENDENT SET

To analyze the running time of the above branching algorithm, let $P(n)$ be the maximum number of leaves in the search tree recursively generated by the algorithm to solve the problem on a graph with n vertices. The worst case recurrence is obtained by the branching on a vertex of degree 3:

$$P(n) \leq P(n-1) + P(n-4).$$

The unique positive real root of the corresponding polynomial $x^4 - x^3 - 1 = 0$ is 1.3802.... Since each recursive call takes polynomial time, and the total number of subproblems solved is within a polynomial factor of $P(n)$, the running time of the algorithm (according to simple analysis) is $\mathcal{O}(1.3803^n)$. Furthermore, let $P_h(n)$, $h \le n$, be the number of subproblems being graphs with h vertices solved when the algorithm solves MIS on a graph with n vertices. By basically the same analysis, one obtains $P_h(n) = \mathcal{O}(1.3803^{n-h})$ for any fixed h.

Now the running time of algorithm mis5 can be reduced by modifying the algorithm with the memorization technique, at the cost of exponential space complexity, in the following way. Whenever we solve a subproblem G', which is an induced subgraph of G, we store the pair $(G', \mathrm{mis}(G'))$ in a database, where $\mathrm{mis}(G')$ is a maximum independent set of G'. Before solving any subproblem, we check whether its solution is already available in the database. Observe that, since G has 2^n induced subgraphs, the database can easily be implemented so that each query takes time polynomial in n.

Let us analyze the running time of the algorithm obtained by memorization. Due to memorization, no subproblem is solved twice. We shall use a balancing argument depending on the value of h. (Note that this is done to analyze the running time. There is no explicit use of h in the algorithm.)

There are $\binom{n}{h}$ induced subgraphs of G with h vertices, which implies $P_h(n) \le \binom{n}{h}$. Moreover the upper bound $P_h(n) = O(1.3803^{n-h})$ still holds. Altogether

$$P_h(n) = O(\min\{1.3803^{n-h}, \binom{n}{h}\}).$$

By setting $h = \alpha n$ and using the binary entropy function and Lemma 3.13, and balancing the two terms, one obtains that, for each h, $P_h(n) \le 1.3803^{n-\alpha n} < 1.3426^n$, where $\alpha > 0.0865$ satisfies

$$1.3803^{1-\alpha} = \frac{1}{\alpha^\alpha (1-\alpha)^{1-\alpha}}$$

Theorem 10.7. *The running time of the exponential space algorithm obtained by applying memorization to algorithm* mis5 *is* $\mathcal{O}(1.3426^n)$.

In the above algorithm the memorization technique was applied to a branching algorithm for which the running time was achieved by simple analysis. It is natural to ask whether memorization can also be combined with a Measure and Conquer analysis. Indeed this is possible and the approach is quite similar, though there is a subtle difference to be taken into account.

To illustrate how to apply memorization to a branching algorithm for which the running time was achieved by Measure and Conquer analysis we reconsider the branching algorithm for the MINIMUM SET COVER algorithm presented in Chap. 6. We rely on the results of the Measure and Conquer analysis of algorithm msc to apply memorization.

Let us recall some important features of the Measure and Conquer analysis of algorithm $\mathtt{msc}$ of Chap. 6. The measure $k = k(\mathcal{S})$ of an instance $\mathcal{S}$ of the MINIMUM SET COVER problem (MSC) is

$$k(\mathcal{S}) = \sum_{i \geq 1} w_i n_i + \sum_{j \geq 1} v_j m_j,$$

where n_i is the number of subsets $S \in \mathcal{S}$ of cardinality i and m_j is the number of elements $u \in \mathcal{U}$ of frequency j. Using the following weights

$$w_i = \begin{cases} 0.377443 & \text{if } i = 2, \\ 0.754886 & \text{if } i = 3, \\ 0.909444 & \text{if } i = 4, \\ 0.976388 & \text{if } i = 5, \end{cases} \quad \text{and} \quad v_i = \begin{cases} 0.399418 & \text{if } i = 2, \\ 0.767579 & \text{if } i = 3, \\ 0.929850 & \text{if } i = 4, \\ 0.985614 & \text{if } i = 5, \end{cases}$$

the Measure and Conquer analysis in Chap. 6 established a running time of $\mathcal{O}(1.2353^{|\mathcal{U}|+|\mathcal{S}|})$ for algorithm $\mathtt{msc}$.

Now let us explain how memorization can be applied to the branching algorithm $\mathtt{msc}$. Let $\mathcal{P}$ be the collection of those subproblems generated during the execution of the algorithm on which the algorithm branches. In particular, none of these subproblems contains a set of cardinality one nor an element of frequency one. Let $P_h(k)$ be the maximum number of subproblems of $\mathcal{P}$ of size h, $0 \leq h \leq k$. By basically the same analysis as in in Sect. 6.3, one obtains $P_h(k) \leq 1.2353^{k-h} \leq 1.2353^{k'-h}$, where $k' := |\mathcal{S}| + |\mathcal{U}|$.

Now we need to discuss the crucial difference between simple analysis and Measure and Conquer when it comes to memorization. In simple analysis the measure of the input is the number of vertices k, and exactly this is used in memorization when subproblems are stored and analysed by cardinality k. When relying on a Measure and Conquer analysis we need to convert the measure h of a subproblem S' into the cardinality i of the set S' to be stored in the database. To do this let us consider a subproblem of size h. Observe that it can be encoded via a pair $(\mathcal{S}', \mathcal{U}')$, where $\mathcal{S}' \subseteq \mathcal{S}$ and $\mathcal{U}' \subseteq \mathcal{U}$. Since the problem considered does not contain any set of cardinality one nor any element of frequency one and by the monotonicity of the weights, we have that each set and each element of the subproblem has weight at least $\min\{v_2, w_2\}$. Consequently

$$|\mathcal{S}'| + |\mathcal{U}'| \leq \lfloor h/\min\{\alpha_2, \beta_2\} \rfloor = \lfloor h/0.377443 \rfloor =: h'.$$

Having translated measure into cardinality we may proceed as in the case of simple analysis.

Since due to the memorization no subproblem is solved more than once, $P_h(k)$ is also upper bounded by

$$P_h(k) \leq \sum_{i \leq h'} \binom{k'}{i}.$$

Observe that, the number of different weights being a constant, the number of possible distinct feasible values of h is a polynomial in k. Putting things together,

$$|\mathcal{P}| \leq \sum_h \min\left\{1.2353^{k'-h}, \sum_{i \leq h'} \binom{k'}{i}\right\}$$

$$= O^*\left(\sum_{h' > k'/2} 1.2353^{k'-h'\min\{v_2,w_2\}} + \sum_{h' \leq k'/2} \min\left\{1.2353^{k'-h'\min\{v_2,w_2\}}, \binom{k'}{h'}\right\}\right)$$

$$= O^*\left(1.1871^{k'} + \max_{h' \leq k'/2} \min\left\{1.2353^{k'-h'\min\{v_2,w_2\}}, \binom{k'}{h'}\right\}\right).$$

Applying Stirling's formula,

$$\max_{h' \leq k'/2} \min\left\{1.2353^{k'-h'\min\{v_2,w_2\}}, \binom{k'}{h'}\right\} = O(1.2353^{k'-0.01996k'}) = O(1.2302^{k'}).$$

Hence, $|\mathcal{P}| = O(1.2302^{k'})$. At each branching step the algorithm removes at least one set. Thus the total number of subproblems is $O^*(|\mathcal{P}|)$. Moreover, the cost of each query to the database is polynomial in k.

Theorem 10.8. *The exponential space algorithm for the* MINIMUM SET COVER *problem established by applying memorization to algorithm* msc *has running time* $O(1.2302^{(|\mathcal{S}|+|\mathcal{U}|)})$. *The corresponding exponential space algorithm for the* MINIMUM DOMINATING SET *problem has a running time of* $O(1.2302^{2n}) = O(1.5132^n)$.

Notes

Savitch's proof that non-deterministic polynomial space is equal to deterministic polynomial space appeared in [195]. The Divide & Conquer approach for the TRAVELLING SALESMAN PROBLEM was proposed by Gurevich and Shelah in 1987 [108]. As was shown by Bodlaender, Fomin, Koster, Kratsch and Thilikos [31] and Björklund and Husfeldt [25], this approach can be used for many other problems.

The algorithm for graph coloring is due to Björklund and Husfeldt [25]. A polynomial space graph coloring algorithm of running time $O(5.283^n)$ is due to Bodlaender and Kratsch [32]. Theorem 10.3 is due to Koivisto and Parviainen [139]. It is stated in a much more general setting than presented in this book.

The first use of memorization in branching algorithms is due to Robson [185]. This technique is now a standard tool in speeding up branching algorithms [36, 84, 212, 87, 143]. Another technique to speedup branching algorithms by using exponential space is based on treewidth and is used in [80].

Some open problems on exponential space requirements of exact algorithms are discussed in [221].

Chapter 11
Miscellaneous

In this chapter we collect several unrelated results. The algorithm solving the BAND-WIDTH MINIMIZATION problem can be seen as a combination of the branching and the dynamic programming techniques. The second section on Branch & Recharge provides a new angle on branching algorithms. The third section gives a brief overview of fundamental results by Impagliazzo, Paturi, and Zane, and the Exponential Time Hypothesis.

11.1 Bandwidth

Let us recall that the objective of the BANDWIDTH MINIMIZATION problem is to find an optimal layout of a graph $G = (V,E)$, i.e. a bijection $f : V \to \{1,\dots,n\}$ such that the width of the layout f,

$$\max_{\{u,v\}\in E} |f(u) - f(v)|$$

is minimized. The layout f can be seen as an embedding of the vertices of G into slots numbered from 1 to n. Thus the bandwidth of a graph is at most B, if there is an embedding in $\{1,2,\dots,n\}$ such that the stretch $|f(u) - f(v)|$ of each edge $\{u,v\}$ of G is at most B. Thus sometimes we will refer to integers $i \in \{1,2,\dots,n\}$ as *slots*.

Obviously, the bandwidth of any graph G on n vertices is at most $n-1$. In this section, we discuss an algorithm that for any B, $1 \le B \le n-1$, decides in time $\mathcal{O}^*(6^n)$ whether the bandwidth of G is at most B. Then running this algorithm for all values of B, the BANDWIDTH MINIMIZATION problem can be solved in time $\mathcal{O}^*(6^n)$.

Let G be a graph on n vertices and B, $1 \le B \le n-1$, be an integer. Let us assume that n is divisible by $B+1$ (this assumption does not change the result but makes the description slightly easier). If G is not connected, then its bandwidth is equal to the maximum bandwidth of its connected components. Thus we may assume that G is

F.V. Fomin, D. Kratsch, *Exact Exponential Algorithms*, Texts in Theoretical
Computer Science. An EATCS Series, DOI 10.1007/978-3-642-16533-7_11,
© Springer-Verlag Berlin Heidelberg 2010

a connected graph. We partition the set of slots $\{1, 2, \ldots, n\}$ into segments

$$\{1, \ldots, B+1\}, \{B+2, \ldots, 2B+2\}, \ldots, \{n-B, \ldots, n\}.$$

With every slot $i \in \{1, 2, \ldots, n\}$ we associate two values

$$x(i) = ((i-1) \mod (B+1)) + 1$$

and

$$y(i) = \lceil \frac{i}{B+1} \rceil.$$

In other words, $x(i)$ is the position of i in its segment, and $y(i)$ is the number of the segment to which i belongs.

For example, for $B = 2$ and $n = 9$, we have the following partition into 3 segments

$$1\,2\,3\,|\,4\,5\,6\,|\,7\,8\,9.$$

In this case, $x(1) = 1$, $y(1) = 1$, $x(6) = 3$, $y(5) = 2$, etc.

We say that the permutation

$$\pi_1\ \pi_2\ \cdots\ \pi_n$$

of slots is the *lexicographical* ordering of slots if for $i \leq j$,

$$(x(\pi_i), y(\pi_i)) \preceq (x(\pi_j), y(\pi_j)).$$

For our example with set of slots $\{1, 2, \ldots, 9\}$ and $B = 2$, the lexicographical ordering of slots is

$$1\,4\,7\,2\,5\,8\,3\,6\,9.$$

Now we are ready to describe the algorithm. The algorithm proceeds in two phases. In the first phase it creates $\mathcal{O}^*(3^n)$ subproblems and in the second phase it solves each of the subproblems in time $\mathcal{O}^*(2^n)$.

The goal of the first phase is to break a possible layout into $\mathcal{O}^*(3^n)$ subproblems, such that for each of the subproblems the approximate position of every vertex is known. More precisely: The first phase starts by computing all possible assignments of the vertices of the graph to the segments. The crucial observation is that for every layout of width at most B, every pair of adjacent vertices is either assigned to the same segment or to adjacent ones. Indeed, if two adjacent vertices are placed into different non adjacent segments, then the stretch of the edge connecting these vertices is at least $B + 2$. Thus if a vertex of a graph is assigned to a segment with number i, then its neighbors can only be placed in one of the segments with numbers $i - 1$, i, or $i + 1$. The number of such assignments satisfying the condition that adjacent vertices of G are assigned to the same or to adjacent segments is at most $\frac{n}{B+1} \cdot 3^n = \mathcal{O}^*(3^n)$. This can be seen as follows. We select (arbitrarily) a vertex v in G and we consider all $n/(B+1)$ possibilities of assigning v to different segments. Then we iteratively select an unassigned vertex u which has a neighbor w that already has been assigned to some segment. If no such vertex u exists then all vertices have been assigned to segments as G is connected. Since u and w are adjacent, there are only three possible segments to which u can be assigned. For each such an assignment we also perform a check leaving only those assignments σ assigning exactly $B+1$ vertices to each of the segments. We also remove assignments containing edges with endpoints in non-adjacent segments. Assignments generated in the first phase will be called *segment assignments*.

To describe the second phase of the algorithm we need the following definition. For a fixed segment assignment $\sigma : V \to \{1, 2, \ldots, n/(B+1)\}$, we say that a subset $A \subseteq V$ is *lexicographically embeddible* in $\{1, 2, \ldots, n\}$ if

a) for each edge $\{u, v\} \in E$, conditions $u \in A$ and $v \notin A$ yield that $\sigma(v) \leq \sigma(u)$. In other words, v can be assigned either to the same segment as u, or to the adjacent segment preceding the slot containing u;

b) there is a mapping $\gamma : A \to \{1, 2, \ldots, n\}$ agreeing with σ, which means that for every vertex $v \in A$, $\gamma(v)$ belongs to the segment containing $\sigma(v)$, and such that the slots occupied by $\gamma(v)$, $v \in A$, are the first $|A|$ slots in the lexicographical ordering of slots.

In Fig. 11.1, we give examples of lexicographically embeddible sets. For graph G in Fig. 11.1 and $B = 2$, the mapping $\sigma(a) = \sigma(b) = \sigma(f) = 1$, $\sigma(c) = \sigma(g) = \sigma(h) = 2$, and $\sigma(d) = \sigma(e) = \sigma(i) = 3$, is the segment assignment. The layout γ agrees with σ and its width is at most B. The sets $\emptyset \subset \{a\} \subset \{a, g\} \subset \{a, d, g\} \subset \{a, d, f, g\} \subset \{a, c, d, f, g\} \subset \{a, c, d, f, g, i\} \subset \{a, b, c, d, f, g, i\} \subset \{a, b, c, d, f, g, h, i\} \subset \{a, b, c, d, e, f, g, h, i\}$ are lexicographically embeddible. The set $P = \{a, c, d, g\}$ is not lexicographically embeddible because the first four slots in the lexicographical ordering are $1, 4, 7, 2$ but every mapping agreeing with σ must use two slots from the second segment $\{4, 5, 6\}$ for c and g. Thus there is no mapping satisfying condition $b)$ of the definition of lexicographically embeddible sets. The set $Q = \{a, b, g, h\}$ is also not lexicographically embeddible because $c \notin Q$ but $\sigma(c) > \sigma(b)$ and thus σ does not satisfy condition $a)$.

Lemma 11.1. *Let σ be a segment assignment. Then the following are equivalent*

- *There is layout $\gamma : V \to \{1, 2, \ldots, n\}$ agreeing with σ and of width at most B*
- *There is sequence $\emptyset = A_0 \subset A_1 \subset \cdots \subset A_n = V$ such that for each $i \in \{1, 2, \ldots, n\}$, $|A_i| = i$, and A_i is lexicographically embeddible in $\{1, 2, \ldots, n\}$.*

Proof. Let γ be a layout of width at most B agreeing with σ. We define the set A_i, $1 \leq i \leq n$, as the set of vertices occupying the first i slots in the lexicographical ordering of slots. Let $u \in A_i$, $v \notin A_i$, be a pair of adjacent vertices, and let k be the slot occupied by u and ℓ be the slot occupied by v in the layout γ. Layout γ agrees with σ, and thus $y(k) = \sigma(u)$, $y(\ell) = \sigma(v)$. The vertices of A_i occupy the first i slots in the lexicographical ordering, and thus in every segment with number at least $\sigma(u)$, the first $y(k) - 1$ slots are occupied by vertices of A_i. Therefore, there are at least B slots between k and the first unoccupied slot of the segment $\sigma(u) + 1$. Since

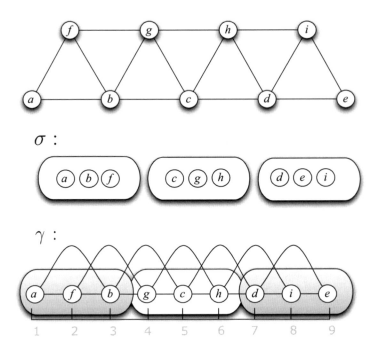

Fig. 11.1 A Graph, its segment assignment and the corresponding embedding

$|k - \ell| \leq B$, we conclude that v can be assigned only to segment $\sigma(u)$ or $\sigma(u) - 1$, and thus $\sigma(v) \leq \sigma(u)$.

For a given sequence $\emptyset = A_0 \subset A_1 \subset \cdots \subset A_n = V$ of sets lexicographically embeddible in $\{1, 2, \ldots, n\}$, we define the layout γ of width at most B as follows. For $v = A_i \setminus A_{i-1}$, we put $\gamma(v) = k$, where k is the ith position in the lexicographical ordering of slots. Let u, v be adjacent vertices with $k = \gamma(u) < \gamma(v) = \ell$. If k and ℓ belong to the same segment, then $\ell - k \leq B$. If $y(k) = y(\ell) - 1$, then there is i, such that $v \in A_i$ and $u \notin A_i$. Hence, $x(k) > x(\ell)$, and thus $\ell - k \leq B$.   □

Now we want to use Lemma 11.1 to compute a layout γ of width at most B which agrees with segment assignment σ. For every vertex subset $A \subseteq V$, we decide whether A is lexicographically embeddible in $\{1, 2, \ldots, n\}$ by making use of dynamic programming.

Every set $A = \{v\}$ of cardinality one is lexicographically embeddible in $\{1, 2, \ldots, n\}$ if and only if σ assigns v to the first segment and all neighbors of v are also assigned to the first segment. This check can clearly be performed in polynomial time. Let A be a set of cardinality at least two that is lexicographically embeddible and let γ be the corresponding mapping. Then for the vertex $v \in A$ assigned by γ to the slot with maximum (in the lexicographical ordering of occupied slots) position, the set $A \setminus \{v\}$ is also lexicographically embeddible. Hence a set A of cardinality at least two is lexicographically embeddible in $\{1, 2, \ldots, n\}$ if and only if there is $v \in A$ such that

- $A \setminus \{v\}$ is also lexicographically embeddible;
- Every vertex $u \notin A$ adjacent to v, is assigned to a slot with number $\sigma(v)$ or $\sigma(v) - 1$.

By making use of this observation, we perform dynamic programming over all subsets of V to compute all lexicographically embeddible subsets of V. This computation requires $\mathcal{O}^*(2^n)$ steps and space. Finally, if the set V is lexicographically embeddible, $V = A_n \supset A_{n-1} \supset \cdots \supset A_1 \supset A_0 = \emptyset$. Then by Lemma 11.1, these sets can be used to construct a layout of width at most B.

Since we try $\mathcal{O}^*(3^n)$ assignments, we have that in total the running time of the algorithm is $\mathcal{O}^*(2^n \cdot 3^n) = \mathcal{O}^*(6^n)$ and we conclude with the following theorem.

Theorem 11.2. *The* BANDWIDTH MINIMIZATION *problem is solvable in time* $\mathcal{O}^*(6^n)$ *using exponential space.*

Exercise 11.3. Improve the running time of the algorithm of Theorem 11.2 to $\mathcal{O}^*(5^n)$.

11.2 Branch & Recharge

In Chap. 6 a Measure & Conquer analysis of branching algorithms has been presented. In this section a branching algorithm is established which solves an enumeration problem for generalized domination. To analyse the running time of the algorithms weights are attributed to the vertices of the graph and these weights are redistributed over the graph by a recharging mechanism. This recharging is used to guarantee that all branchings have essentially the same branching vector, which leads to an easy time analysis. This approach allows us to construct and analyse a branching algorithm for a class of problems while the branching algorithms in previous chapters are problem dependent.

(σ, ρ)-*Dominating Set Enumeration Problem.* Let σ and ρ be two nonempty sets of nonnegative integers. In the (σ, ρ)-DOMINATING SET ENUMERATION PROBLEM (Enum-(σ, ρ)-DS) we are given an undirected graph $G = (V, E)$. The task is to enumerate all (σ, ρ)-dominating sets $S \subseteq V$ of G, i.e. S is a vertex subset satisfying that $|N(v) \cap S| \in \sigma$ for all $v \in S$ and $|N(v) \cap S| \in \rho$ for all $v \in V \setminus S$.

The table in Fig. 11.2 shows a few previously defined and studied graph invariants which can be expressed in this framework.

We consider the Enum-(σ, ρ)-DS problem assuming that σ is successor-free, i.e. does not contain a pair of consecutive integers, and that both σ and ρ are finite. These conditions are satisfied, for example, for perfect codes and strong stable sets (see Fig. 11.2).

σ	ρ	(σ,ρ)-dominating set
$\{0,1,2,\ldots\}$	$\{1,2,3,\ldots\}$	dominating set
$\{0\}$	$\{0,1,2,\ldots\}$	independent set
$\{0\}$	$\{1\}$	perfect code
$\{0\}$	$\{0,1\}$	strong stable set
$\{0\}$	$\{1,2,3,\ldots\}$	independent dominating set
$\{1\}$	$\{1\}$	total perfect dominating set

Fig. 11.2 Examples of (σ,ρ)-dominating sets

We present an algorithm enumerating all (σ,ρ)-dominating sets using a Branch & Recharge approach. The upper bound of the running time of the branching algorithm immediately implies a combinatorial upper bound on the number of (σ,ρ)-dominating sets.

Theorem 11.4. *If σ is successor-free and both sets σ and ρ are finite then all the (σ,ρ)-dominating sets of an input graph $G = (V,E)$ without isolated vertices can be enumerated in time $\mathcal{O}^*(c^n)$, where $c = c_{p,q} < 2$ is a constant depending only on $p = \max\sigma$ and $q = \max\rho$. Moreover, every isolate-free graph G contains $\mathcal{O}^*(c^n)$ (σ,ρ)-dominating sets, where $c = c_{p,q} < 2$ (is the same constant).*

To see that one may assume $\max\{p,q\} > 0$, notice that if $\sigma = \rho = \{0\}$ then for every isolate-free graph the empty set is the unique (σ,ρ)-dominating set.

Now we describe the details of the branching algorithm. The recursive algorithm `enum-sigma-rho` described in Fig. 11.3 is the principal part of the branching algorithm solving `Enum-`(σ,ρ)`-DS` for an input graph $G = (V,E)$. The overall enumeration algorithm first computes an (arbitrary) breadth-first search (BFS) ordering $v_1,v_2,\ldots,v_n$ of the input graph G. Then we call `enum-sigma-rho`$(G,\emptyset,\emptyset,[])$ where $[]$ is an empty list. This algorithm outputs a list L containing all (σ,ρ)-dominating sets. (The list is organized in such a way that it does not contain multiples and also concatenation does not create multiples.) A simple check of the output list L removes all sets S which are not (σ,ρ)-dominating sets and outputs the list of all (σ,ρ)-dominating sets.

Similar to Measure & Conquer analysis of graph algorithms, weights are used to analyse the branching algorithm. To analyse the branching algorithm `enum-sigma-rho` one first assigns a weight of 1 to each vertex of the input graph. Thus the input graph has (total) weight $w(G) = \sum_{v \in V} w(v) = n$. In every instance of a subproblem $(G,S,\overline{S},L)$, the weight of a vertex of the graph G is a real in $[0,1]$. Furthermore the total weight of the graph of a subproblem is smaller than the weight of the original problem. Finally if a vertex is assigned to S or $\overline{S}$ then its weight is set to 0. This set-up is similar to the analyses in Chaps. 2 and 6.

Now the new idea of the analysis is to guarantee the branching vector $(1,1+\varepsilon)$ for all branchings, where $\varepsilon = \frac{1}{\max\{p,q\}} > 0$ and $p = \max\sigma$, $q = \max\rho$. Because of $\varepsilon > 0$ this immediately implies a running time of $\mathcal{O}^*(c^n)$, where $c < 2$ is the positive real root of the characteristic polynomial $x^{1+\varepsilon} - x^\varepsilon - 1$.

Algorithm enum-sigma-rho$(G, S, \overline{S}, L)$.
Input: A graph $G = (V, E)$, disjoint vertex subsets S, $\overline{S}$, and a list L of candidate vertex subsets.
Output: List L of all (σ, ρ)-dominating sets D in G satisfying $S \subseteq D$ and $\overline{S} \cap D = \emptyset$.

> **if** *there is no free vertex* **then** $L := L \cup \{S\}$
> **else**
>> let v be the last free vertex in the BFS ordering of V
>> **if** v *is the first vertex in the BFS ordering* **then**
>>> $L := L \cup \{S, S \cup \{v\}\}$
>>> Halt
>>
>> **else**
>>> **if** $\exists$ *free vertex* x *s.t.* x *adjacent to* v *and* $|N(x) \cap S| = \max\{p, q\}$ **then**
>>>> $\overline{S} := \overline{S} \cup \{v\}$
>>>
>>> **else**
>>>> **if** $\exists x \in S$ *s.t.* v *is its unique free neighbor* **then**
>>>>> **if** $|N(x) \cap S| \in \sigma$ **then** $\overline{S} := \overline{S} \cup \{v\}$
>>>>> **if** $|N(x) \cap S| + 1 \in \sigma$ **then** $S := S \cup \{v\}$
>>>>> **if** $\{|N(x) \cap S|, |N(x) \cap S| + 1\} \cap \sigma = \emptyset$ **then** Halt
>>
>> **if** $\exists x$ *s.t.* $|N(x) \cap S| > \max\{p, q\}$ **then** Halt
>> **if** v *is (still) free* **then**
>>> concatenate the lists enum-sigma-rho$(G, S, \overline{S} \cup \{v\}, L)$ and
>>> enum-sigma-rho$(G, S \cup \{v\}, \overline{S}, L)$
>>
>> **else** enum-sigma-rho$(G, S, \overline{S}, L)$

Fig. 11.3 Algorithm enum-sigma-rho computes a list L containing all (σ, ρ)-dominating sets of a graph G (and possibly some more vertex sets) when called for G, $S = \emptyset$, $\overline{S} = \emptyset$ and empty list L.

How can this be achieved? Firstly algorithm enum-sigma-rho contains only one branching rule. Suppose the algorithm branches on the free vertex v and $w(v) < 1$. Recharging guarantees that the weight of v is increased such that $w(v) = 1$ before the branching on vertex v is performed. How recharging is done will be explained later. If the algorithm branches on a vertex v, then vertex v is either *selected*, i.e. chosen in the solution $S \subseteq V$, or *discarded*, i.e. not chosen in the solution $S \subseteq V$ and thus taken into $\overline{S}$. When branching on a vertex v, guaranteed to have weight 1, then the weight of the graph decreases by at least 1 when v is discarded, and it decreases by at least $1 + \varepsilon$ when v is selected in the so far generated candidate S for a (σ, ρ)-dominating set. In each case the weight of G decreases by 1 since the weight of v is set to 0. In case of selection a free neighbor of v sends a weight of ε, i.e. its weight is decreased by ε, and the weight of G is decreased by $1 + \varepsilon$.

It remains to describe how the weight of the vertex v, chosen to branch on since it is the last free vertex in the BFS ordering of G, can be reset to 1 if $w(v) > 1$. Let us emphasize that this recharging is done without changing the weight of the graph. The procedure recharge is described in Fig. 11.4. Notice that this procedure takes as input the vertex v whose weight has to be increased to 1, the current weight function $w : V \to [0, 1]$ of the graph G and a directed graph H on the same vertex set as G storing all exchanges of weights such that there is a directed edge from u to v in H iff u sent a weight of ε to v, in the execution producing the current instance.

Procedure recharge$(v, S, \overline{S}, w, H)$

if $w(v) < 1$ **then**
 let $\{w_1, \ldots, w_t\} = \{x : \{v, x\} \in E(H)\}$
 for $i := 1$ **to** t **do** let u_i be another free neighbor (in G) of w_i
 for $i := 1$ **to** t **do**
  $w(u_i) := w(u_i) - \varepsilon$
  $E(H) := (E(H) \cup \{\{u_i, w_i\}\}) \setminus \{\{v, w_i\}\}$
 $w(v) := 1$

Fig. 11.4 Procedure recharge

Note that in the procedure recharge the vertices $w_1, \ldots, w_t$ are distinct, while $u_1, \ldots, u_t$ need not be. If some u is the chosen free neighbor of several, say k, vertices from $w_1, \ldots, w_t$, then its weight drops by $k\varepsilon$ and also k edges starting at u are added to H. It can be shown that each w_i has another free neighbor in G. To guarantee that recharging works correctly and is always possible it is crucial that σ is successor-free. The process of recharging a vertex v is illustrated in Fig. 11.5.

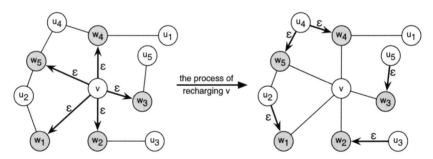

Fig. 11.5 Recharging vertex v

On the left side of Fig. 11.5, a vertex v sent a weight of ε to each of its neighbors w_1, w_2, w_3, w_4 and w_5, at the time when w_i was assigned to S. Since σ is successor-free, each w_i, $i \in \{1, \ldots, 5\}$, has a free neighbor u_k, $k \in \{1, \ldots, 5\}$ (otherwise, v would be forced by a reduction rule and thus the algorithm would not branch on v). The value of ε ensures that each free vertex u_k has a weight no smaller than ε (otherwise, such a vertex would have more neighbors in S than allowed by σ and ρ). On the right side of Fig. 11.5, the vertex v is recharged. For each k, a weight of ε is sent from u_k to one or more w_i. The redistribution of the weights ensures $w(v) = 1$.

The correctness of the recharging procedure and the fact that the weight of the vertex v to branch on can always be reset to 1 by redistribution of weights (without changing the weight of the graph) can be shown by a careful analysis of the distribution of the weights on the vertices of the graph and its changes. For more details and proofs we refer to the original paper [81].

Finally let us show that now the running time of the branching algorithm can be established easily. Recall that the weight of an instance $(G, w, S, \overline{S}, H, L)$ is $w(G) = \sum_{v \in V} w(v)$. In each branching on a vertex v the (total) weight of the graph G decreases by 1 when discarding v, and it decreases by $1 + \varepsilon$ when selecting v. Hence the only branching rule of the algorithm has branching vector $(1, 1+\varepsilon)$. The running time of each execution of `enum-sigma-rho` (without recursive calls) is polynomial, and so the total running time of the algorithm is $\mathcal{O}^*(T)$ where T is the number of leaves of the search tree. Note that each (σ, ρ)-dominating set corresponds to one leaf of the search tree.

Let $T[k]$ be the maximum number of leaves of the search tree that any execution of our algorithm may generate on a problem instance of weight k. Due to the branching vector we obtain:

$$T[k] \leq T[k-1] + T[k-1-\varepsilon].$$

Thus the number of (σ, ρ)-dominating sets (which is bounded from above by $T[n]$) in an isolate-free graph on n vertices is at most c^n, and the running time of our algorithm that enumerates all of them is $\mathcal{O}^*(c^n)$, where c is the largest real root of the characteristic polynomial

$$x^{1+\varepsilon} - x^{\varepsilon} - 1.$$

The table shows the base of the exponential function bounding the running time of our algorithm for some particular values of $\varphi = \max\{p, q\}$.

φ	1	2	3	4	5	6	7	100
c	1.6181	1.7549	1.8192	1.8567	1.8813	1.8987	1.9116	1.9932

Exercise 11.5. An efficient dominating set of a graph G is a (σ, ρ)-dominating set with $\sigma = \{0\}$ and $\rho = \{1\}$. The ENUM-$(\{0\}, \{1\})$-DS problem can be solved in time $\mathcal{O}(1.6181^n)$ as shown above. Construct and analyse a (significantly) faster branching algorithm solving this enumeration problem.

11.3 Subexponential Algorithms and ETH

The hypothesis $P \neq NP$ implies only that no NP-complete problem can be solved in polynomial time. However this does not exclude the possibility of solving, say, MAXIMUM INDEPENDENT SET (MIS) in time $2^{\mathcal{O}(\sqrt{n})}$. For most of the algorithms we have discussed so far, the worst case upper bound on the running time was of order $2^{\mathcal{O}(n)}$ and it is natural to ask whether such behaviour is common for all NP-complete problems. Of course not. For example, it is possible to show that the treewidth of a planar graph on n vertices is $\mathcal{O}(\sqrt{n})$. Combining this fact with the algorithms on graphs of bounded treewidth from Chap. 5, we obtain that many problems like MAXIMUM INDEPENDENT SET or MINIMUM DOMINATING SET which are still NP-complete on planar graphs, are solvable on these graphs in time $\mathcal{O}(2^{\sqrt{n}})$. However for many natural NP-complete problems on general classes of graphs, sets, or formulas, we do not know subexponential algorithms.

Before proceeding further, we need to define precisely what we mean by a subexponential algorithm. We say that a function $f : \mathbb{R} \to \mathbb{R}$ is a *subexponential function*, if $f(n) \in 2^{o(n)}$. In other words, for every $\varepsilon > 0$ there is n_0 such that for every $n \geq n_0$, $|f(n)| \leq 2^{\varepsilon n}$.

We should be careful when defining subexponential algorithms. It seems natural to define a subexponential time algorithm as an algorithm that for any input I of length $|I|$ of a problem Π outputs the correct answer in time $2^{o(|I|)}$ or in other words, in time subexponential of the input length. However there is a problem with such definition.

Let us illustrate the problem with the following example. The MAXIMUM CLIQUE problem is to find a clique of maximum size in a given graph G. This problem is equivalent to MIS: G has a clique of size k if and only if its complement $\overline{G}$ has an independent set of size k. We do not know any algorithm solving MIS in time subexponential in the number of vertices of the graph, and thus we know no such algorithm for solving MAXIMUM CLIQUE. However, the following lemma shows that if we parameterize the running time of the algorithm for MAXIMUM CLIQUE by the number of edges, then there is a subexponential time algorithm for MAXIMUM CLIQUE.

Lemma 11.6. *The* MAXIMUM CLIQUE *problem on graphs with m edges is solvable in time* $2^{\mathcal{O}(\sqrt{m})}$.

Proof. Let us assume that G is a connected graph. We choose a vertex v of minimum degree. If the degree of v is at least $\sqrt{2m}$, then because

$$2m = \sum_{v \in V} |N(v)| \geq n\sqrt{2m},$$

we have that $n \leq \sqrt{2m}$. Then the running time of the brute force algorithm trying all possible vertex subsets is $n2^{\mathcal{O}(n)} = n2^{\mathcal{O}(\sqrt{m})}$.

If the degree of v is less than $\sqrt{2m}$, let us consider the following simple branching algorithm finding a clique C of maximum size in G. Pick a vertex v of minimum degree and branch on two subproblems, in one v is a vertex of C and in the other subproblem v is not a vertex of C. In the first subproblem, we use brute force to search the neighbourhood of v for a clique of maximum size and thus branch on all possible subsets of $N(v)$. In the second case we call the algorithm recursively.

We argue by induction on m that the number of steps of the algorithm is at most $n2^{\sqrt{2m}}$. For $m = 0$, G has only one clique and the size of this clique is 1. To solve the subproblem corresponding to the case $v \in C$, we use brute force to select a clique C' of maximum size in $N(v)$. Then $C = C' \cup \{v\}$. Since $|N(v)| < \sqrt{2m}$, there are at most $2^{\sqrt{2m}}$ steps in this case. If $v \notin C$, then by the induction assumption, the problem is solvable in at most $(n-1)2^{\sqrt{2m}}$ steps. Thus the total number of steps of the algorithm is

$$2^{\sqrt{2m}} + (n-1)2^{\sqrt{2m}} = n2^{\sqrt{2m}}$$

and its running time is $2^{\mathcal{O}(\sqrt{m})}$. $\square$

The conclusion here is that every time we speak about subexponential algorithms, we also should mention what is the corresponding parameterization. An interesting fact that will be proved below is that for many problems the existence of a subexponential algorithm parameterized by the number of vertices, by the number of edges, by the number of variables, or by the number of clauses is "equivalently unlikely".

First we explain what we mean by "unlikely". In the world of polynomial algorithms we have a widely believed hypothesis $P \neq NP$, which is equivalent to the hypothesis that 3-SAT is not solvable in polynomial time. Despite many attempts, there is no known algorithm solving 3-SAT with n variables in time $2^{o(n)}$. The following hypothesis is stronger than the $P \neq NP$ hypothesis.

Exponential Time Hypothesis (ETH): *There is no algorithm solving 3-SAT in time $2^{o(n)}$, where n is the number of variables of the input CNF formula.*

With the current state of the art in exact algorithms, ETH seems to be very reasonable. Can we prove that subject to ETH there are other problems that cannot be solved in subexponential time? The difficulty here is that almost all known classical polynomial time reductions cannot be used to prove this type of statement. For example, the standard reduction to MIS maps an instance of 3-SAT with n variables and m clauses to a graph with $\mathcal{O}(n+m)$ vertices. This graph can have $\Omega(n^3)$ vertices and this type of reduction only shows that MIS is not solvable in time $2^{o(\sqrt[3]{n})}$. This reduction would be subexponential time preserving if we parameterize the problems by clauses and edges but it is not subexponential with a more natural parameterization by variables and vertices.

We want to prove that some problems are equivalent subject to solvability in subexponential time. The idea is that since we are not restricted by polynomial time anymore, we can use subexponential time reductions between problems. First we need the following simple technical lemma.

Lemma 11.7. *Let $f : [0, +\infty) \to \mathbb{R}$ be a non-negative non-decreasing function. The following are equivalent*

i) $f(n) \in 2^{o(n)}$;
ii) *There is a function $h : \mathbb{R} \to \mathbb{R}$, such that for every $\varepsilon > 0$, $f(n) \leq h(\varepsilon) \cdot 2^{\varepsilon n}$.*

Proof. i) $\Rightarrow$ *ii).* For every $\varepsilon > 0$, there is n_ε such that for every $n \geq n_\varepsilon$, $f(n) \leq 2^{\varepsilon n}$. We define $h(\varepsilon) = 2^{n_\varepsilon}$. Then $f(n) \leq h(\varepsilon) \cdot 2^{\varepsilon n}$. Indeed, for $n \geq n_\varepsilon$, $f(n) \leq 2^{\varepsilon n}$ and because $f(n)$ is a non-decreasing, $f(n) \leq 2^{n_\varepsilon} = h(\varepsilon)$ for $n \leq n_\varepsilon$.

ii) $\Rightarrow$ *i).* For $\varepsilon > 0$, let us choose $\delta = \varepsilon/2$ and $n_0 \geq \log_2 h(\delta)/\delta$. Then because $f(n) \leq h(\delta) \cdot 2^{\delta n}$, we have that for every $n > n_0$, $h(\delta) \cdot 2^{\delta n} \leq 2^{\delta n_0} \cdot 2^{\delta n} \leq 2^{\varepsilon n}$. $\square$

Theorem 11.8. *For every $k \geq 3$, k-SAT can be solved in time $2^{o(n)}$ if and only if it can be solved in time $2^{o(m)}$, where n is the number of variables and m is the number of clauses of the input k-CNF formula F.*

In the heart of the proof of Theorem 11.8 lies the Sparsification Lemma of Impagliazzo, Paturi, and Zane [120]. The proof of the Sparsification Lemma is based on a clever branching on variables contained in many clauses. The proof of the lemma is quite technical and we do not give it here.

Lemma 11.9. *(Sparsification Lemma) Let $k \geq 2$. There is a computable function $g : \mathbb{N} \to \mathbb{N}$ such that for every $\ell \in N$ and every formula $F \in k$-CNF with n variables, there is a formula $F' = \bigvee_{i \in [t]} F_i$ such that:*

- *F' is equivalent to F;*
- *$t \leq 2^{n/\ell}$;*
- *each F_i is a subformula of F in which each variable appears at most $g(\ell)$ times.*

Furthermore there is an algorithm that, given F and ℓ, computes F' in time $2^{n/\ell} |F|^{O(1)}$.

Now using Lemmata 11.7 and 11.9, we prove Theorem 11.8.

Proof (Proof of Theorem 11.8). Suppose that there is an algorithm solving k-SAT on instances with m clauses and n variables in time $2^{o(n)}$. Because every instance is an k-CNF, $m \geq n/k$. Hence $2^{o(n)} = 2^{o(m)}$ for every fixed k.

For the other direction, let us assume that k-SAT is solvable in time $2^{o(m)}$. We want to use the Sparsification Lemma to show that for every fixed ℓ, k-CNF can be solved in time $\mathcal{O}(2^{n/\ell} p(n+m))$ for every $\ell \in \mathbb{N}$, where p is a polynomial. This together with Lemma 11.7 will imply that k-CNF can be solved in time $2^{o(n)} p(n + m)$. Because for every fixed $k \geq 3$, $2^{o(n)} p(n+m) = 2^{o(n)}$, this will complete the proof of the theorem.

Since k-SAT can be solved in time $2^{o(m)}$, by Lemma 11.7, we have that it can be solved in time $\mathcal{O}(h(s)2^{m/s})$ for every $s \in \mathbb{N}$. Let $\ell \in \mathbb{N}$ and F be a k-CNF formula with m clauses and n variables. We apply the Sparsification Lemma with $\ell' = 2\ell$ and obtain in time $2^{n/\ell'} |F|^{O(1)}$ an equivalent formula $F' = \bigvee_{i \in [t]} F_i$ with $t \leq 2^{n/\ell'}$ and at most $n \cdot g(\ell')$ clauses. Now we solve each of the subformulas F_i using an algorithm for k-SAT running in time $h(s)2^{m/s} |F_i|^{O(1)}$ for every $s \in \mathbb{N}$. We choose $s = \ell' \cdot g(\ell')$, and then the total running time of the algorithm is

$$T(n+m) = 2^{n/\ell'} |F|^{O(1)} + \sum_{i \in [t]} h(\ell' \cdot g(\ell'))2^{\frac{n \cdot g(\ell')}{\ell' \cdot g(\ell')}} |F_i|^{O(1)}$$
$$\leq 2^{n/\ell'} |F|^{O(1)} + 2^{n/\ell'} h(\ell' \cdot g(\ell'))2^{\frac{n}{\ell'}} |F|^{O(1)}.$$

The fact that $|F_i|^{O(1)} \leq |F|^{O(1)}$ follows from the fact that F_i is a subformula of F. By putting $h'(\ell) = h(\ell' \cdot g(\ell'))$, we have that

$$T(n+m) \leq h'(\ell)2^{n/\ell}(n+m)^{O(1)},$$

and thus by Lemma 11.7, $T(n+m) = 2^{o(n)}p(n+m)$ for some polynomial p. This concludes the proof of the theorem. □

By making use of Theorem 11.8, one can prove that many problems do not have subexponential algorithms unless ETH fails. Let us prove the following result as an example.

Corollary 11.10. *Unless ETH fails, there is no algorithm solving* MIS *on instances with n vertices and m edges in time $2^{o(n)}$ or $2^{o(m)}$.*

Proof. Without loss of generality we can assume that the input graph G is connected. Thus $n \leq m - 1$. We take the following standard example of polynomial time many-one (or Karp) reduction from 3-SAT to MIS (see, e.g. Chap. 7 in Sipser's textbook [206]) which for a given 3-CNF formula F with n variables and m clauses constructs a graph G with $2n + m$ vertices and $\mathcal{O}(n+m)$ edges such that F is satisfiable if and only if G has an independent set of size $n + m$.

An algorithm finding a maximum independent set in G in time $2^{o(n)}$ also works in time $2^{o(m)}$ and thus can be used to decide in time $2^{o(m)}$ whether F is satisfiable. But by Theorem 11.8, this yields that 3-SAT is solvable in time $2^{o(n)}$ and thus ETH fails.

Impagliazzo, Paturi, and Zane [120] developed a theory which can be used to show that a problem does not have subxponential time algorithm unless ETH fails. The important notion in this theory is the notion of *SERF* reducibility. The idea behind SERF reducibility is that to preserve subexponential time it is important that the relevant parameters do not increase more than linearly but the time complexity of the reduction is less important. Let P_1 be a problem with complexity parameter m_1 and P_2 be a problem with complexity parameter m_2. For a many-one reduction f from P_1 to P_2 to preserve subexponential time, we want $m_2(f(x)) = O(m_1(x))$, where $f(x)$ is an instance of P_2 obtained from an instance x of P_1. This is also sometimes called *strong many-one reduction*. Many of the standard reductions between NP-complete problems are of this nature at least for some right choice of complexity parameters. For an example most of the reductions from k-SAT are of this nature if we use the number of clauses as our complexity parameters, but *not* if we use the number of variables. This is exactly the point where Theorem 11.8 becomes so important. Another feature is that we allow the reduction to take subexponential time. Now we formally define a subexponential reduction family (SERF).

A subexponential reduction family (SERF) is a collection of Turing reductions M_ε from P_1 to P_2 such that for each $\varepsilon > 0$:

1. $M_\varepsilon(x)$ runs in time poly$(|x|) \, 2^{\varepsilon m_1(x)}$;
2. If $M_\varepsilon(x)$ queries P_2 with input x' then $m_2(x') = O(m_1(x))$ and $|x| = |x'|^{O(1)}$.

If such a reduction family exists we say that P_1 is SERF-reducible to P_2. The Strong many to one reduction is a special case of SERF-reducibility.

In this terminology, the Sparsification Lemma shows that k-SAT with the number of variables as the complexity parameter is SERF-reducible to k-SAT with the number of clauses as the complexity parameter. Many classical reductions are strong many-one reductions, and therefore many problems including k-COLORING (finding a colouring of a graph in k colours), MINIMUM VERTEX COVER, MAXIMUM CLIQUE, and HAMILTONIAN PATH parameterized by the number of vertices, or k-MINIMUM SET COVER (find a minimum covering of a set by sets of size at most k) parameterized by the number of elements do not have subexponential algorithms unless ETH fails.

Notes

The first algorithm for BANDWIDTH MINIMIZATION breaking the trivial $\mathcal{O}^*(n!)$ bound is attributed to Feige and Kilian in [75]. The running time of the Feige-Kilian algorithm is $\mathcal{O}^*(10^n)$. The algorithm presented in this chapter is due to Cygan and Pilipczuk [55]. Fomin, Lokhstanov and Saurabh used a similar approach to obtain an exact algorithm to compute an embedding of a graph into the line with minimum distortion [92]. The fastest algorithm for the BANDWIDTH MINIMIZATION problem has running time $\mathcal{O}(4.473^n)$ [56].

GENERALIZED DOMINATION was introduced by Telle [214]. Fomin, Golovach, Kratochvil, Kratsch and Liedloff presented a Branch & Recharge algorithm to enumerate the (σ, ρ)-dominating sets in [81]. The successor-freeness of σ is crucial for the Branch & Recharge algorithm presented. It is an interesting question whether it is possible to develop a different recharging mechanism that does not need the successor-freeness of σ.

Lemma 11.6 is from the paper of Stearns and Hunt [208]. The Exponential Time Hypothesis (ETH) is due to Impagliazzo, Paturi, and Zane [120]. In [120] ETH is stated in a slightly weaker form. For $k \geq 3$, Impagliazzo, Paturi, and Zane define the numbers

$$s_k = \inf\{\delta : \text{there is a } 2^{\delta n} \text{ algorithm for solving } k\text{-SAT}\}$$

and

(ETH): *For every fixed $k \geq 3$, $s_k > 0$.*

Impagliazzo and Paturi show that s_k increases infinitely often assuming ETH [119]. ETH discussed in this chapter is a stronger assumption than ETH of Impagliazzo, Paturi, and Zane. The differences between the two hypotheses is that for ETH to fail ours requires the existence of a uniform algorithm solving k-SAT in subexponential time, while the definition of Impagliazzo, Paturi, and Zane does not exclude the possibility that for different δ different algorithms are selected. The version of ETH in this chapter is commonly used now, see e.g. the book of Flum and Grohe [78]. ETH establishes interesting links between exact algorithms and Parameterized Complexity (see Chap. 16 of Flum and Grohe [78]).

The Sparsification Lemma and SERF-reducibility are due to Impagliazzo, Paturi, and Zane [120]. Independently, Johnson and Szegedy obtained a sparsification result for MIS [126]. An improvement of the Sparsification Lemma is given by Calabro, Impagliazzo and Paturi in [45].

Another, much stronger hypothesis, is that $s_k \to 1$ when $k \to \infty$. Pătraşcu and Williams [171] provide some evidence for this hypothesis.

Despite many attempts, the best known deterministic algorithms for SAT run in time

$$2^{n\left(1 - \frac{1}{\mathcal{O}(\log(m/n))}\right)} \cdot poly(m),$$

see, e.g. [60].

Chapter 12
Conclusions, Open Problems and Further Directions

We conclude with a number of open problems. Some of them are of a fundamental nature and some of them can serve as starting points for newcomers in the field.

Fundamental questions

Every problem in NP can be solved by enumerating all solution candidates. The question is whether such trivial enumeration can be avoided for every problem in NP. In other words, is brute-force search the only approach to solve NP problems in general? A positive answer to this question implies that P $\neq$ NP. On the other hand, the assumption P $\neq$ NP does not yield a negative answer. Recent work of Williams demonstrates that: "... carrying out the seemingly modest program of finding slightly better algorithms for all search problems may be extremely difficult (if not impossible)" [219].

Most of the exact algorithms are problem dependent—almost every specific problem requires specific arguments to show that this problem can be solved faster than brute-force search. In the world of polynomial time algorithms and parameterized complexity we possess very powerful tools allowing us to establish efficient criteria to identify large classes of polynomial time solvable or fixed parameter tractable problems. It would be desirable to obtain generic tools allowing us to identify large classes of NP-complete problems solvable faster than by brute-force search.

Every algorithmic theory becomes fruitful when accompanied by complexity theory. In the current situation we are only able to distinguish between exponential and subexponential running times (subject to Exponential Time Hypotheses). A challenge here is to develop a theory of exponential lower bounds. For example, is it possible to prove (up to some plausible assumption from complexity like P$\neq$NP, FPT$\neq$W[1], ETH, etc.) that there is no algorithm solving 3-SAT on n variables in time 1.000000001^n?

More concrete questions

Three fundamental NP-complete problems, namely, SAT, TSP and GRAPH COLORING can be solved within the same running time $\mathcal{O}^*(2^n)$. Obtaining for any of

F.V. Fomin, D. Kratsch, *Exact Exponential Algorithms*, Texts in Theoretical Computer Science. An EATCS Series, DOI 10.1007/978-3-642-16533-7_12, © Springer-Verlag Berlin Heidelberg 2010 these problems an algorithm of running time $\mathcal{O}^*((2-\varepsilon)^n)$ for any $\varepsilon > 0$ would be exciting.

Can it be that for every $\varepsilon > 0$ the existence of an $\mathcal{O}^*((2-\varepsilon)^n)$ algorithm for one of these three problems yields an $O((2-\delta)^n)$ algorithm for the other two, for some $\delta > 0$? Recently, Björklund [22] announced a randomized algorithm solving HAMILTONIAN PATH in time $\mathcal{O}(1.66^n)$.

Many permutation and partition problems can be solved in time $\mathcal{O}^*(2^n)$ by dynamic programming which requires exponential space. An interesting question is whether there are $\mathcal{O}^*(2^n)$ time and polynomial space algorithms for TSP, GRAPH COLORING, and TREEWIDTH.

Some permutation problems like PATHWIDTH or TREEWIDTH can be solved in time $\mathcal{O}^*((2-\varepsilon)^n)$ (and exponential space). What can we say about DIRECTED FEEDBACK ARC SET, CUTWIDTH and HAMILTONIAN CYCLE?

The running time of current branching algorithms for MIS with more and more detailed analyses seems to converge somewhere near $\mathcal{O}^*(1.2^n)$. It appears that obtaining an algorithm running in time $\mathcal{O}^*(1.1^n)$ will require completely new ideas. Similarly the question can be asked whether MDS is solvable in time $\mathcal{O}(1.3^n)$. MINIMUM DIRECTED FEEDBACK VERTEX SET requires us to remove the minimum number of vertices of a directed graph such that the remaining graph is acyclic. The problem is trivially solvable in time $\mathcal{O}^*(2^n)$. The trivial algorithm was beaten by Razgon with an algorithm running in $\mathcal{O}(1.9977^n)$ time [178]. It seems that improving even to $\mathcal{O}^*(1.8^n)$ is a difficult problem.

SUBGRAPH ISOMORPHISM is trivially solvable in time $\mathcal{O}^*(2^{n \log n})$. Is it possible to solve this problem in time $2^{\mathcal{O}(n)}$? A similar question can be asked about GRAPH HOMOMORPHISM. In CHROMATIC INDEX (also known as EDGE COLORING) the task is to color edges with the minimum number of colors such that no two edges of the same color are incident. The only non-trivial algorithm we are aware of reduces the problem to (vertex) graph coloring of the line graph. This takes time $\mathcal{O}^*(2^m)$. Is CHROMATIC INDEX solvable in time $2^{\mathcal{O}(n)}$?

Enumerating the number of certain objects is a fundamental question in combinatorics. Sometimes such questions can be answered using exact algorithms. Consider the following general problem: "For a given property π, what is the maximum number of vertex subsets with property π in a graph on n vertices?" For example, the theorem of Moon-Moser says that when the property π is "being a maximal clique", then this number is $3^{n/3}$. But for many other natural properties, we still do not know precise (even asymptotically) bounds. For example, for minimal dominating sets the correct value is between 1.5704^n and 1.7159^n [88], for minimal feedback vertex sets between 1.5926^n and 1.7548^n [79]. For minimal feedback vertex sets in tournaments the old bounds of Moon [160]—1.4757^n and 1.7170^n—were recently improved by Gaspers and Mnich to 1.5448^n and 1.6740^n [102]. For minimal separators we know that the number is between 1.4423^n and 1.6181^n [95], for potential maximal cliques between 1.4423^n and 1.7347^n [96].

References

1. Achlioptas, D., Beame, P., Molloy, M.: Exponential bounds for DPLL below the satisfiability threshold. In: Proceedings of the 15th Annual ACM-SIAM Symposium on Discrete Algorithms (SODA 2004), pp. 139–140. SIAM (2004)
2. Aigner, M.: A course in enumeration, *Graduate Texts in Mathematics*, vol. 238. Springer, Berlin (2007)
3. Alekhnovich, M., Hirsch, E.A., Itsykson, D.: Exponential lower bounds for the running time of DPLL algorithms on satisfiable formulas. J. Automat. Reason. **35**(1-3), 51–72 (2005).
4. Alon, N., Spencer, J.: The Probabilistic Method, third edn. John Wiley (2008)
5. Amini, O., Fomin, F.V., Saurabh, S.: Counting subgraphs via homomorphisms. In: Proceedings of the 36th International Colloquium on Automata, Languages and Programming (ICALP 2009), *Lecture Notes in Comput. Sci.*, vol. 5555, pp. 71–82. Springer (2009)
6. Angelsmark, O., Jonsson, P.: Improved algorithms for counting solutions in constraint satisfaction problems. In: Proceedings of the 9th International Conference on Principles and Practice of Constraint Programming (CP 2003), *Lecture Notes in Comput. Sci.*, vol. 2833, pp. 81–95. Springer (2003)
7. Arnborg, S., Proskurowski, A.: Linear time algorithms for NP-hard problems restricted to partial k-trees. Disc. Appl. Math. **23**(1), 11–24 (1989)
8. Aspvall, B., Plass, M.F., Tarjan, R.E.: A linear-time algorithm for testing the truth of certain quantified boolean formulas. Inf. Process. Lett. **8**(3), 121–123 (1979)
9. Babai, L., Kantor, W.M., Luks, E.M.: Computational complexity and the classification of finite simple groups. In: Proceedings of the 24th Annual Symposium on Foundations of Computer Science (FOCS 1983), pp. 162–171. IEEE (1983)
10. Bax, E., Franklin, J.: A finite-difference sieve to count paths and cycles by length. Inf. Process. Lett. **60**(4), 171–176 (1996)
11. Bax, E.T.: Inclusion and exclusion algorithm for the Hamiltonian path problem. Inf. Process. Lett. **47**(4), 203–207 (1993)
12. Bax, E.T.: Algorithms to count paths and cycles. Inf. Process. Lett. **52**(5), 249–252 (1994)
13. Beigel, R.: Finding maximum independent sets in sparse and general graphs. In: Proceedings of the 10th ACM-SIAM Symposium on Discrete Algorithms (SODA 1999), pp. 856–857. SIAM (1999)
14. Beigel, R., Eppstein, D.: 3-coloring in time $O(1.3289^n)$. Journal of Algorithms **54**(2), 168–204 (2005)
15. Bellman, R.: Dynamic programming. Princeton University Press, Princeton, N. J. (1957)
16. Bellman, R.: Combinatorial processes and dynamic programming. In: Proc. Sympos. Appl. Math., Vol. 10, pp. 217–249. American Mathematical Society, Providence, R.I. (1960)
17. Bellman, R.: Dynamic programming treatment of the travelling salesman problem. J. ACM **9**, 61–63 (1962)

F.V. Fomin, D. Kratsch, *Exact Exponential Algorithms*, Texts in Theoretical
Computer Science. An EATCS Series, DOI 10.1007/978-3-642-16533-7,
© Springer-Verlag Berlin Heidelberg 2010

18. Berge, C.: Graphs and hypergraphs, *North-Holland Mathematical Library*, vol. 6. North-Holland Publishing Co., Amsterdam (1973)

19. Berry, A., Bordat, J.P., Cogis, O.: Generating all the minimal separators of a graph. Int. J. Found. Comput. Sci. **11**(3), 397–403 (2000)

20. Bezrukov, S., Elsässer, R., Monien, B., Preis, R., Tillich, J.P.: New spectral lower bounds on the bisection width of graphs. Theor. Comp. Sci. **320**(2-3), 155–174 (2004)

21. Biere, A., Heule, M.J.H., van Maaren, H., Walsh, T. (eds.): Handbook of Satisfiability, *Frontiers in Artificial Intelligence and Applications*, vol. 185. IOS Press (2009)

22. Björklund, A.: Determinant sums for undirected hamiltonicity. In: Proceedings of the 51st Annual IEEE Symposium on Foundations of Computer Science (FOCS 2010), to appear. IEEE (2010)

23. Björklund, A.: Exact covers via determinants. In: Proceedings of the 27th International Symposium on Theoretical Aspects of Computer Science (STACS 2010), *Leibniz International Proceedings in Informatics (LIPIcs)*, vol. 5, pp. 95–106. Schloss Dagstuhl–Leibniz-Zentrum fuer Informatik, Dagstuhl, Germany (2010).

24. Björklund, A., Husfeldt, T.: Inclusion-exclusion algorithms for counting set partitions. In: Proceedings of the 47th Annual IEEE Symposium on Foundations of Computer Science (FOCS 2006), pp. 575–582. IEEE (2006)

25. Björklund, A., Husfeldt, T.: Exact algorithms for exact satisfiability and number of perfect matchings. Algorithmica **52**(2), 226–249 (2008)

26. Björklund, A., Husfeldt, T., Kaski, P., Koivisto, M.: Fourier meets Möbius: Fast subset convolution. In: Proceedings of the 39th Annual ACM Symposium on Theory of Computing (STOC 2007), pp. 67–74. ACM (2007)

27. Björklund, A., Husfeldt, T., Kaski, P., Koivisto, M.: Computing the Tutte polynomial in vertex-exponential time. In: Proceedings of the 49th Annual IEEE Symposium on Foundations of Computer Science (FOCS 2008), pp. 677–686. IEEE (2008).

28. Björklund, A., Husfeldt, T., Kaski, P., Koivisto, M.: The travelling salesman problem in bounded degree graphs. In: Proceedings of the 35th International Colloquium on Automata, Languages and Programming (ICALP 2008), *Lecture Notes in Comput. Sci.*, vol. 5125, pp. 198–209. Springer (2008)

29. Björklund, A., Husfeldt, T., Kaski, P., Koivisto, M.: Trimmed Moebius inversion and graphs of bounded degree. In: Proceedings of the 25th Annual Symposium on Theoretical Aspects of Computer Science (STACS 2008), *Dagstuhl Seminar Proceedings*, vol. 08001, pp. 85–96. Internationales Begegnungs- und Forschungszentrum fuer Informatik (IBFI), Schloss Dagstuhl, Germany (2008)

30. Björklund, A., Husfeldt, T., Koivisto, M.: Set partitioning via inclusion–exclusion. SIAM J. Comput. **39**(2), 546–563 (2009).

31. Bodlaender, H.L., Fomin, F.V., Koster, A.M.C.A., Kratsch, D., Thilikos, D.M.: On exact algorithms for treewidth. In: Proceedings of the 14th Annual European Symposium on Algorithms (ESA 2006), *Lecture Notes in Comput. Sci.*, vol. 4168, pp. 672–683. Springer (2006)

32. Bodlaender, H.L., Kratsch, D.: An exact algorithm for graph coloring with polynomial memory. Technical Report UU-CS-2006-015, University of Utrecht (March 2006)

33. Bondy, J.A., Murty, U.S.R.: Graph theory, *Graduate Texts in Mathematics*, vol. 244. Springer, New York (2008).

34. Bouchitté, V., Todinca, I.: Treewidth and minimum fill-in: Grouping the minimal separators. SIAM J. Comput. **31**(1), 212–232 (2001)

35. Bouchitté, V., Todinca, I.: Listing all potential maximal cliques of a graph. Theor. Comput. Sci. **276**(1-2), 17–32 (2002)

36. Bourgeois, N., Croce, F.D., Escoffier, B., Paschos, V.T.: Exact algorithms for dominating clique problems. In: Proceedings of the 20th International Symposium on Algorithms and Computation (ISAAC 2009), *Lecture Notes in Comput. Sci.*, vol. 5878, pp. 4–13. Springer (2009)

37. Bourgeois, N., Escoffier, B., Paschos, V.T.: An $O^*(1.0977^n)$ exact algorithm for max independent set in sparse graphs. In: Proceedings of the 3rd Workshop on Parameterized and

Exact Computation (IWPEC 2008), *Lecture Notes in Comput. Sci.*, vol. 5018, pp. 55–65. Springer (2008)

38. Bourgeois, N., Escoffier, B., Paschos, V.T., van Rooij, J.M.M.: A bottom-up method and fast algorithms for max independent set. In: Proceedings of the 12th Scandinavian Symposium and Workshops on Algorithm Theory (SWAT 2010), *Lecture Notes in Comput. Sci.*, vol. 6139, pp. 62–73. Springer (2010)

39. Bourgeois, N., Escoffier, B., Paschos, V.T., van Rooij, J.M.M.: Maximum independent set in graphs of average degree at most three in $O(1.08537^n)$ average degree. In: Proceedings of the 7th Annual Conference on Theory and Applications of Models of Computation (TAMC 2010), *Lecture Notes in Comput. Sci.*, vol. 6108, pp. 373–384. Springer (2010)

40. Broersma, H., Fomin, F.V., van 't Hof, P., Paulusma, D.: Fast exact algorithms for hamiltonicity in claw-free graphs. In: Proceedings of the 35th International Workshop on Graph-Theoretic Concepts in Computer Science (WG 2009), *Lecture Notes in Comput. Sci.*, vol. 5911, pp. 44–53 (2009)

41. Brueggemann, T., Kern, W.: An improved deterministic local search algorithm for 3-SAT. Theor. Comp. Sci. **329**(1-3), 303–313 (2004)

42. Byskov, J.M.: Enumerating maximal independent sets with applications to graph colouring. Operations Research Letters **32**(6), 547–556 (2004)

43. Byskov, J.M.: Exact algorithms for graph colouring and exact satisfiability. Ph.D. thesis, University of Aarhus, Denmark (2004)

44. Byskov, J.M., Madsen, B.A., Skjernaa, B.: New algorithms for exact satisfiability. Theoret. Comput. Sci. **332**(1-3), 515–541 (2005).

45. Calabro, C., Impagliazzo, R., Paturi, R.: A duality between clause width and clause density for SAT. Proceedings of the 21st Annual IEEE Conference on Computational Complexity (CCC 2006) pp. 252–260 (2006).

46. Cameron, P.J.: Combinatorics: topics, techniques, algorithms. Cambridge University Press, Cambridge (1994)

47. Chen, J., Kanj, I.A., Jia, W.: Vertex cover: further observations and further improvements. Journal of Algorithms **41**(2), 280–301 (2001)

48. Chen, J., Kanj, I.A., Xia, G.: Labeled search trees and amortized analysis: improved upper bounds for NP-hard problems. Algorithmica **43**(4), 245–273 (2005)

49. Chung, F.R.K., Graham, R.L., Frankl, P., Shearer, J.B.: Some intersection theorems for ordered sets and graphs. J. Combin. Theory Ser. A **43**(1), 23–37 (1986)

50. Chvátal, V.: Determining the stability number of a graph. SIAM J. Comput. **6**(4), 643–662 (1977)

51. Coppersmith, D., Winograd, S.: Matrix multiplication via arithmetic progressions. J. Symbolic Comput. **9**(3), 251–280 (1990)

52. Cormen, T.H., Leiserson, C., Rivest, R., Stein, C.: Introduction to Algorithms, second edn. The MIT Press, Cambridge, Mass. (2001)

53. Courcelle, B.: The monadic second-order logic of graphs I: Recognizable sets of finite graphs. Information and Computation **85**, 12–75 (1990)

54. Courcelle, B.: The monadic second-order logic of graphs III: Treewidth, forbidden minors and complexity issues. Informatique Théorique **26**, 257–286 (1992)

55. Cygan, M., Pilipczuk, M.: Faster exact bandwidth. In: Proceedings of the 34th International Workshop on Graph-Theoretic Concepts in Computer Science (WG 2008), *Lecture Notes in Comput. Sci.*, vol. 5344, pp. 101–109. Springer (2008)

56. Cygan, M., Pilipczuk, M.: Exact and approximate bandwidth. In: Proceedings of the 36th International Colloquium on Automata, Languages and Programming (ICALP 2009), *Lecture Notes in Comput. Sci.*, vol. 5555, pp. 304–315. Springer (2009)

57. Dahllöf, V., Jonsson, P., Beigel, R.: Algorithms for four variants of the exact satisfiability problem. Theoret. Comput. Sci. **320**(2-3), 373–394 (2004).

58. Dahllöf, V., Jonsson, P., Wahlström, M.: Counting models for 2SAT and 3SAT formulae. Theor. Comp. Sci. **332**(1-3), 265–291 (2005)

59. Dantsin, E., Goerdt, A., Hirsch, E.A., Kannan, R., Kleinberg, J., Papadimitriou, C., Ragha-van, P., Schöning, U.: A deterministic $(2 - 2/(k + 1))^n$ algorithm for k-SAT based on local search. Theor. Comp. Sci. **289**(1), 69–83 (2002)

60. Dantsin, E., Hirsch, E.A.: Worst-case upper bounds. In: Handbook of Satisfiability, *Frontiers in Artificial Intelligence and Applications*, vol. 185, chap. 12, pp. 403–424. IOS Press (2009)

61. Dasgupta, S., Papadimitriou, C., Vazirani, U.: Algorithms. McGraw-Hill (2008)

62. Davis, M., Logemann, G., Loveland, D.: A machine program for theorem-proving. Comm. ACM **5**, 394–397 (1962)

63. Davis, M., Putnam, H.: A computing procedure for quantification theory. J. ACM **7**, 201–215 (1960)

64. Díaz, J., Serna, M., Thilikos, D.M.: Counting H-colorings of partial k-trees. Theor. Comput. Sci. **281**, 291–309 (2002)

65. Diestel, R.: Graph theory, *Graduate Texts in Mathematics*, vol. 173, third edn. Springer-Verlag, Berlin (2005)

66. Downey, R.G., Fellows, M.R.: Parameterized complexity. Springer-Verlag, New York (1999)

67. Eisenbrand, F., Grandoni, F.: On the complexity of fixed parameter clique and dominating set. Theor. Comput. Sci. **326**(1-3), 57–67 (2004)

68. Ellis, J.A., Sudborough, I.H., Turner, J.S.: The vertex separation and search number of a graph. Information and Computation **113**(1), 50–79 (1994)

69. Eppstein, D.: Small maximal independent sets and faster exact graph coloring. In: Proceedings of the 7th Workshop on Algorithms and Data Structures (WADS 2001), *Lecture Notes in Comput. Sci.*, vol. 2125, pp. 462–470. Springer, Berlin (2001)

70. Eppstein, D.: Small maximal independent sets and faster exact graph coloring. Journal of Graph Algorithms and Applications **7**(2), 131–140 (2003)

71. Eppstein, D.: Quasiconvex analysis of backtracking algorithms. In: Proceedings of the 15th ACM-SIAM Symposium on Discrete Algorithms (SODA 2004), pp. 781–790. SIAM (2004)

72. Eppstein, D.: Quasiconvex analysis of multivariate recurrence equations for backtracking algorithms. ACM Trans. Algorithms **2**(4), 492–509 (2006).

73. Eppstein, D.: The traveling salesman problem for cubic graphs. Journal of Graph Algorithms and Applications **11**(1), 61–81 (2007)

74. Feder T., Motwani, R.: Worst-case time bounds for coloring and satisfiability problems. J. Algorithms **45**(2), 192-201 (2002).

75. Feige, U.: Coping with the NP-hardness of the graph bandwidth problem. In: Proceedings of the 7th Scandinavian Workshop on Algorithm Theory (SWAT 2000), *Lecture Notes in Comput. Sci.*, vol. 1851, pp. 10–19. Springer (2000)

76. Feller, W.: An introduction to probability theory and its applications. Vol. I, third edn. John Wiley & Sons Inc., New York (1968)

77. Fernau, H., Kneis, J., Kratsch, D., Langer, A., Liedloff, M., Raible, D., Rossmanith, P.: An exact algorithm for the maximum leaf spanning tree problem. In: Proceedings of the 4th International Workshop on Parameterized and Exact Computation (IWPEC 2009), *Lecture Notes in Comput. Sci.*, vol. 5917, pp. 161–172. Springer (2009)

78. Flum, J., Grohe, M.: Parameterized Complexity Theory. Texts in Theoretical Computer Science. An EATCS Series. Springer-Verlag, Berlin (2006)

79. Fomin, F.V., Gaspers, S., Pyatkin, A.V., Razgon, I.: On the minimum feedback vertex set problem: Exact and enumeration algorithms. Algorithmica **52**(2), 293–307 (2008).

80. Fomin, F.V., Gaspers, S., Saurabh, S., Stepanov, A.A.: On two techniques of combining branching and treewidth. Algorithmica **54**(2), 181–207 (2009)

81. Fomin, F.V., Golovach, P.A., Kratochvíl, J., Kratsch, D., Liedloff, M.: Branch and recharge: Exact algorithms for generalized domination. In: Proceedings of the 10th Workshop on Algorithms and Data Structures, (WADS 2007), *Lecture Notes in Comput. Sci.*, vol. 4619, pp. 507–518. Springer (2007)

82. Fomin, F.V., Golovach, P.A., Kratochvíl, J., Kratsch, D., Liedloff, M.: Sort and search: Exact algorithms for generalized domination. Inf. Process. Lett. **109**(14), 795–798 (2009)

83. Fomin, F.V., Grandoni, F., Kratsch, D.: Measure and conquer: Domination – a case study. In: Proceedings of the 32nd International Colloquium on Automata, Languages and Programming (ICALP 2005), *Lecture Notes in Comput. Sci.*, vol. 3580, pp. 191–203. Springer (2005)

84. Fomin, F.V., Grandoni, F., Kratsch, D.: Some new techniques in design and analysis of exact (exponential) algorithms. Bulletin of the EATCS **87**, 47–77 (2005)

85. Fomin, F.V., Grandoni, F., Kratsch, D.: Measure and conquer: A simple $O(2^{0.288n})$ independent set algorithm. In: Proceedings of the 17th Annual ACM-SIAM Symposium on Discrete Algorithms (SODA 2006), pp. 18–25. SIAM (2006)

86. Fomin, F.V., Grandoni, F., Kratsch, D.: Solving connected dominating set faster than 2^n. Algorithmica **52**(2), 153–166 (2008)

87. Fomin, F.V., Grandoni, F., Kratsch, D.: A measure & conquer approach for the analysis of exact algorithms. J. ACM **56**(5) (2009)

88. Fomin, F.V., Grandoni, F., Pyatkin, A.V., Stepanov, A.: Bounding the number of minimal dominating sets: a measure and conquer approach. In: Proceedings of the 16th Annual International Symposium on Algorithms and Computation (ISAAC 2005), *Lecture Notes in Comput. Sci.*, vol. 3827, pp. 573–582. Springer (2005)

89. Fomin, F.V., Grandoni, F., Pyatkin, A.V., Stepanov, A.A.: Combinatorial bounds via measure and conquer: bounding minimal dominating sets and applications. ACM Trans. Algorithms **5 (1)** (2008). Article 9

90. Fomin, F.V., Kratsch, D., Todinca, I., Villanger, Y.: Exact algorithms for treewidth and minimum fill-in. SIAM J. Comput. **38**(3), 1058–1079 (2008)

91. Fomin, F.V., Kratsch, D., Woeginger, G.J.: Exact (exponential) algorithms for the dominating set problem. In: Proceedings of the 30th Workshop on Graph Theoretic Concepts in Computer Science (WG 2004), *Lecture Notes in Comput. Sci.*, vol. 3353, pp. 245–256. Springer (2004)

92. Fomin, F.V., Lokshtanov, D., Saurabh, S.: An exact algorithm for minimum distortion embedding. In: Proceedings of the 35th International Workshop on Graph-Theoretic Concepts in Computer Science (WG 2009), *Lecture Notes in Comput. Sci.*, vol. 5911, pp. 112–121. Springer (2009)

93. Fomin, F.V., Mazoit, F., Todinca, I.: Computing branchwidth via efficient triangulations and blocks. Disc. Appl. Math. **157**(12), 2726–2736 (2009)

94. Fomin, F.V., Thilikos, D.M.: Branchwidth of graphs. In: M.Y. Kao (ed.) Encyclopedia of Algorithms. Springer (2008)

95. Fomin, F.V., Villanger, Y.: Treewidth computation and extremal combinatorics. In: Proceedings of the 34th International Colloquium on Automata, Languages and Programming (ICALP 2008), *Lecture Notes in Comput. Sci.*, vol. 5125, pp. 210–221. Springer (2008)

96. Fomin, F.V., Villanger, Y.: Finding induced subgraphs via minimal triangulations. In: Proceedings of the 27th International Symposium on Theoretical Aspects of Computer Science (STACS 2010), *Leibniz International Proceedings in Informatics*, vol. 5, pp. 383–394. Schloss Dagstuhl—Leibniz-Zentrum fuer Informatik (2010)

97. Fürer, M.: A faster algorithm for finding maximum independent sets in sparse graphs. In: Proceedings of the 7th Latin American Theoretical Informatics Symposium (LATIN 2006), *Lecture Notes in Comput. Sci.*, vol. 3887, pp. 491–501. Springer (2006)

98. Fürer, M.: Faster integer multiplication. SIAM J. Comput. **39**(3), 979–1005 (2009).

99. Fürer, M., Kasiviswanathan, S.P.: Algorithms for counting 2-SAT solutions and colorings with applications. In: Proceedings of the 3rd International Conference on Algorithmic Aspects in Information and Management (AAIM 2007), *Lecture Notes in Comput. Sci.*, vol. 4508, pp. 47–57. Springer (2007)

100. Garey, M.R., Johnson, D.S.: Computers and Intractability, A Guide to the Theory of NP-Completeness. W.H. Freeman and Company, New York (1979)

101. Gaspers, S.: Exponential time algorithms: Structures, measures, and bounds. Ph.D. thesis, University of Bergen (2008)

102. Gaspers, S., Mnich, M.: On feedback vertex sets in tournaments. In: Proceedings of the 18th European Sympsium on Algorithms (ESA 2010), *Lecture Notes in Comput. Sci.*, vol. 6346, pp. 267–277. Springer (2010)

103. Gaspers, S., Sorkin, G.B.: A universally fastest algorithm for Max 2-Sat, Max 2-CSP, and everything in between. In: Proceedings of the Twentieth Annual ACM-SIAM Symposium on Discrete Algorithms (SODA 2009), pp. 606–615. SIAM (2009)

104. Graham, R.L., Knuth, D.E., Patashnik, O.: Concrete mathematics: A foundation for computer science, second edn. Addison-Wesley Publishing Company, Reading, MA (1994)

105. Gramm, J., Hirsch, E.A., Niedermeier, R., Rossmanith, P.: Worst-case upper bounds for MAX-2-SAT with an application to MAX-CUT. Disc. Appl. Math. **130**(2), 139–155 (2003)

106. Grandoni, F.: Exact algorithms for hard graph problems. Ph.D. thesis, Università di Roma "Tor Vergata" (2004)

107. Gupta, S., Raman, V., Saurabh, S.: Fast exponential algorithms for maximum r-regular induced subgraph problems. In: Proceedings of the 26th International Conference Foundations of Software Technology and Theoretical Computer Science, (FSTTCS 2006), Lecture Notes in Comput. Sci., pp. 139–151. Springer (2006)

108. Gurevich, Y., Shelah, S.: Expected computation time for Hamiltonian path problem. SIAM J. Comput. **16**(3), 486–502 (1987)

109. Halin, R.: S-functions for graphs. J. Geometry **8**(1-2), 171–186 (1976)

110. Hardy, G.H., Ramanujan, S.: Asymptotic formulae in combinatory analysis. Proc. London Math. Soc. **17**, 75–115 (1918)

111. Held, M., Karp, R.M.: A dynamic programming approach to sequencing problems. Journal of SIAM **10**, 196–210 (1962)

112. Hirsch, E.A.: New worst-case upper bounds for SAT. J. Automat. Reason. **24**(4), 397–420 (2000)

113. Hlinený, P., il Oum, S., Seese, D., Gottlob, G.: Width parameters beyond tree-width and their applications. Comput. J. **51**(3), 326–362 (2008)

114. Hochbaum, D.S. (ed.): Approximation algorithms for NP-hard problems. PWS Publishing Co., Boston, MA, USA (1997)

115. Hoffman, A.J., Wolfe, P.: History. In: The traveling salesman problem: A guided tour of combinatorial optimization, Wiley-Intersci. Ser. Discrete Math., pp. 1–15. Wiley, Chichester (1985)

116. Hofmeister, T., Schöning, U., Schuler, R., Watanabe, O.: Randomized algorithms for 3-SAT. Theory Comput. Syst. **40**(3), 249–262 (2007).

117. Horowitz, E., Sahni, S.: Computing partitions with applications to the knapsack problem. J. ACM **21**, 277–292 (1974)

118. Howgrave-Graham, N., Joux, A.: New generic algorithms for hard knapsacks. In: Proceedings of the 29th Annual International Conference on the Theory and Applications of Cryptographic Techniques (EUROCRYPT 2010), *Lecture Notes in Comput. Sci.*, vol. 6110, pp. 235–256. Springer (2010)

119. Impagliazzo, R., Paturi, R.: On the complexity of k-SAT. J. Comput. System Sci. **62**(2), 367–375 (2001).

120. Impagliazzo, R., Paturi, R., Zane, F.: Which problems have strongly exponential complexity? J. Comput. System Sci. **63**(4), 512–530 (2001)

121. Itai, A., Rodeh, M.: Finding a minimum circuit in a graph. SIAM J. Comput. **7**(4), 413–423 (1978)

122. Iwama, K.: Worst-case upper bounds for k-SAT. Bulletin of the EATCS **82**, 61–71 (2004)

123. Iwama, K., Nakashima, T.: An improved exact algorithm for cubic graph TSP. In: 13th Annual International Conference on Computing and Combinatorics (COCOON 2007), *Lecture Notes in Comput. Sci.*, vol. 4598, pp. 108–117. Springer (2007)

124. Iwama, K., Tamaki, S.: Improved upper bounds for 3-SAT. In: Proceedings of the 15th ACM-SIAM Symposium on Discrete Algorithms (SODA 2004), p. 328. SIAM (2004)

125. Jian, T.: An $O(2^{0.304n})$ algorithm for solving maximum independent set problem. IEEE Trans. Computers **35**(9), 847–851 (1986)

126. Johnson, D.S., Szegedy, M.: What are the least tractable instances of max independent set? In: Proceedings of the 10th ACM-SIAM Symposium on Discrete Algorithms (SODA 1999), pp. 927–928. SIAM (1999)

127. Karp, R.M.: Dynamic programming meets the principle of inclusion and exclusion. Oper. Res. Lett. **1**(2), 49–51 (1982)
128. Kawabata, T., Tarui, J.: On complexity of computing the permanent of a rectangular matrix. IECIE Trans. on Fundamentals of Electronics **82**(5), 741–744 (1999)
129. Kleinberg, J., Tardos, E.: Algorithm design. Addison-Wesley, Boston, MA, USA (2005)
130. Kleine Büning, H., Lettman, T.: Propositional logic: deduction and algorithms, *Cambridge Tracts in Theoretical Computer Science*, vol. 48. Cambridge University Press, Cambridge (1999)
131. Klinz, B., Woeginger, G.J.: Faster algorithms for computing power indices in weighted voting games. Math. Social Sci. **49**(1), 111–116 (2005).
132. Kloks, T.: Treewidth, Computations and Approximations, *Lecture Notes in Comput. Sci.*, vol. 842. Springer (1994)
133. Kneis, J., Langer, A., Rossmanith, P.: A fine-grained analysis of a simple independent set algorithm. In: IARCS Annual Conference on Foundations of Software Technology and Theoretical Computer Science (FSTTCS 2009), *Leibniz International Proceedings in Informatics (LIPIcs)*, vol. 4, pp. 287–298. Schloss Dagstuhl–Leibniz-Zentrum fuer Informatik, Dagstuhl, Germany (2009).
134. Kneis, J., Mölle, D., Richter, S., Rossmanith, P.: A bound on the pathwidth of sparse graphs with applications to exact algorithms. SIAM J. Discrete Math. **23**(1), 407–427 (2009).
135. Knuth, D.E.: The art of computer programming, Vol. 3: Seminumerical algorithms, third edn. Addison-Wesley (1998)
136. Kohn, S., Gottlieb, A., Kohn, M.: A generating function approach to the traveling salesman problem. In: Proceedings of the ACM annual conference (ACM 1977), pp. 294–300. ACM Press (1977)
137. Koivisto, M.: An $O(2^n)$ algorithm for graph coloring and other partitioning problems via inclusion-exclusion. In: Proceedings of the 47th Annual IEEE Symposium on Foundations of Computer Science (FOCS 2006), pp. 583–590. IEEE (2006)
138. Koivisto, M.: Partitioning into sets of bounded cardinality. In: Proceedings of the 4th International Workshop on Parameterized and Exact Computation (IWPEC 2009), *Lecture Notes in Comput. Sci.*, vol. 5917, pp. 258–263. Springer (2009)
139. Koivisto, M., Parviainen, P.: A space—time tradeoff for permutation problems. In: Proceedings of the 21th ACM-SIAM Symposium on Discrete Algorithms (SODA 2010), pp. 484–493. ACM and SIAM (2010)
140. Kojevnikov, A., Kulikov, A.S.: A new approach to proving upper bounds for max-2-sat. In: Proceedings of the 17th Annual ACM-SIAM Symposium on Discrete Algorithms (SODA 2006), pp. 11–17. SIAM (2006)
141. Kowalik, L.: Improved edge-coloring with three colors. Theor. Comput. Sci. **410**(38-40), 3733–3742 (2009)
142. Kratochvíl, J., Kratsch, D., Liedloff, M.: Exact algorithms for (2, 1)-labeling of graphs. In: Proceedings of the 32nd International Symposium on Mathematical Foundations of Computer Science (MFCS 2007), *Lecture Notes in Comput. Sci.*, vol. 4708, pp. 513–524. Springer (2007)
143. Kratsch, D., Liedloff, M.: An exact algorithm for the minimum dominating clique problem. Theor. Comput. Sci. **385**(1-3), 226–240 (2007)
144. Kulikov, A.S.: An upper bound $O(2^{0.1625n})$ for exact-3-satisfiability: a simpler proof. Zap. Nauchn. Sem. S.-Peterburg. Otdel. Mat. Inst. Steklov. (POMI) **293**(Teor. Slozhn. Vychisl. 7), 118–128, 183 (2002).
145. Kulikov, A.S., Fedin, S.S.: Solution of the maximum cut problem in time $2^{|E|/4}$. Rossiĭskaya Akademiya Nauk. Sankt-Peterburgskoe Otdelenie. Matematicheskiĭ Institut im. V. A. Steklova. Zapiski Nauchnykh Seminarov (POMI) **293**(Teor. Slozhn. Vychisl. 7), 129–138, 183 (2002)
146. Kulikov, A.S., Kutskov, K.: New upper bounds for the maximum satisfiability problem. Diskret. Mat. **21**(1), 139–157 (2009).
147. Kullmann, O.: New methods for 3-SAT decision and worst-case analysis. Theor. Comp. Sci. **223**(1-2), 1–72 (1999)

148. Kullmann, O.: Fundaments of branching heuristics. In: Handbook of Satisfiability, *Frontiers in Artificial Intelligence and Applications*, vol. 185, chap. 7, pp. 205–244. IOS Press (2009)

149. Lawler, E.L.: A comment on minimum feedback arc sets. IEEE Transactions on Circuits and Systems **11**(2), 296–297 (1964)

150. Lawler, E.L.: A note on the complexity of the chromatic number problem. Inf. Process. Lett. **5**(3), 66–67 (1976)

151. Liedloff, M.: Algorithmes exacts et exponentiels pour les problèmes NP-difficiles : domination, variantes et généralisations. Ph.D. thesis, University of Metz (2007)

152. Liedloff, M.: Finding a dominating set on bipartite graphs. Inf. Process. Lett. **107**(5), 154–157 (2008).

153. van Lint, J.H., Wilson, R.M.: A course in combinatorics, second edn. Cambridge University Press, Cambridge (2001)

154. Lokshtanov, D., Nederlof, J.: Saving space by algebraization. In: Proceedings of 42th ACM Symposium on Theory of Computing (STOC 2010), pp. 321–330. ACM (2010)

155. Miller, R.E., Muller, D.E.: A problem of maximum consistent subsets. IBM Research Rep. RC-240, J.T. Watson Research Center, Yorktown Heights, New York, USA (1960)

156. Minc, H.: Permanents. Cambridge University Press, New York, NY, USA (1984)

157. Mitzenmacher, M., Upfal, E.: Probability and Computing: Randomized Algorithms and Probabilistic Analysis. Cambridge University Press, New York, NY, USA (2005)

158. Monien, B., Preis, R.: Upper bounds on the bisection width of 3- and 4-regular graphs. J. Discrete Algorithms **4**(3), 475–498 (2006)

159. Monien, B., Speckenmeyer, E.: Solving satisfiability in less than 2^n steps. Disc. Appl. Math. **10**(3), 287–295 (1985)

160. Moon, J.W.: On maximal transitive subtournaments. Proc. Edinburgh Math. Soc. (2) **17**, 345–349 (1971)

161. Moon, J.W., Moser, L.: On cliques in graphs. Israel Journal of Mathematics **3**, 23–28 (1965)

162. Nederlof, J.: Fast polynomial-space algorithms using Möbius inversion: Improving on Steiner tree and related problems. In: Proceedings of the 36th International Colloquium on Automata, Languages and Programming (ICALP 2009), *Lecture Notes in Comput. Sci.*, vol. 5555, pp. 713–725. Springer (2009)

163. Nešetřil, J., Poljak, S.: On the complexity of the subgraph problem. Comment. Math. Univ. Carolin. **26**(2), 415–419 (1985)

164. Niedermeier, R.: Invitation to fixed-parameter algorithms, *Oxford Lecture Series in Mathematics and its Applications*, vol. 31. Oxford University Press, Oxford (2006)

165. Nijenhuis, A., Wilf, H.S.: Combinatorial Algorithms. Academic Press, Inc. (1978)

166. Nikolenko, S.I.: Hard satisfiability formulas for DPLL-type algorithms. Zap. Nauchn. Sem. S.-Peterburg. Otdel. Mat. Inst. Steklov. (POMI) **293**(Teor. Slozhn. Vychisl. 7), 139–148, 183 (2002).

167. Oum, S.i.: Computing rank-width exactly. Inf. Process. Lett. **109**(13), 745–748 (2009)

168. Oum, S.i., Seymour, P.: Approximating clique-width and branch-width. J. Combin. Theory Ser. B **96**(4), 514–528 (2006)

169. Papadimitriou, C.H.: On selecting a satisfying truth assignment (extended abstract). In: Proceedings of the 32nd annual Symposium on Foundations of Computer Science (FOCS 1991), pp. 163–169. IEEE (1991).

170. Parsons, T.D.: Pursuit-evasion in a graph. In: Theory and applications of graphs, *Lecture Notes in Math.*, vol. 642, pp. 426–441. Springer, Berlin (1978)

171. Pătraşcu, M., Williams, R.: On the possibility of faster SAT algorithms. In: Proceedings of the 21th ACM-SIAM Symposium on Discrete Algorithms (SODA 2010), pp. 1065–1075. SIAM (2010)

172. Paturi, R., Pudlák, P., Saks, M.E., Zane, F.: An improved exponential-time algorithm for k-SAT. J. ACM **52**(3), 337–364 (2005).

173. Paturi, R., Pudlák, P., Zane, F.: Satisfiability coding lemma. Chicago J. Theor. Comput. Sci. **1999**, Article 11 (1999).

174. Petrov, N.N.: Some extremal search problems on graphs. Differentsial'nye Uravneniya **18**(5), 821–827 (1982)

175. Raible, D., Fernau, H.: A new upper bound for max-2-sat: A graph-theoretic approach. In: Proceedings of the 33rd International Symposium on Mathematical Foundations of Computer Science (MFCS 2008), *Lecture Notes in Comput. Sci.*, vol. 5162, pp. 551–562. Springer (2008)

176. Randerath, B., Schiermeyer, I.: Exact algorithms for MINIMUM DOMINATING SET. Technical Report zaik-469, Zentrum für Angewandte Informatik Köln, Germany (2004)

177. Razgon, I.: Exact computation of maximum induced forest. In: Proceedings of the 10th Scandinavian Workshop on Algorithm Theory (SWAT 2006), *Lecture Notes in Comput. Sci.*, vol. 4059, pp. 160–171. Springer (2006)

178. Razgon, I.: Computing minimum directed feedback vertex set in $O(1.9977^n)$. In: Proceedings of the 10th Italian Conference on Theoretical Computer Science, (ICTCS 2007), pp. 70–81. World Scientific (2007)

179. Razgon, I.: Faster computation of maximum independent set and parameterized vertex cover for graphs with maximum degree 3. J. Discrete Algorithms **7**(2), 191–212 (2009)

180. Riege, T., Rothe, J., Spakowski, H., Yamamoto, M.: An improved exact algorithm for the domatic number problem. Inf. Process. Lett. **101**(3), 101–106 (2007)

181. Robertson, N., Seymour, P.D.: Graph minors. III. Planar tree-width. J. Combin. Theory Ser. B **36**, 49–64 (1984)

182. Robertson, N., Seymour, P.D.: Graph minors. II. Algorithmic aspects of tree-width. Journal of Algorithms **7**(3), 309–322 (1986)

183. Robertson, N., Seymour, P.D.: Graph minors. V. Excluding a planar graph. J. Combin. Theory Ser. B **41**(1), 92–114 (1986)

184. Robertson, N., Seymour, P.D.: Graph minors. X. Obstructions to tree-decomposition. J. Combin. Theory Ser. B **52**(2), 153–190 (1991)

185. Robson, J.M.: Algorithms for maximum independent sets. Journal of Algorithms **7**(3), 425–440 (1986)

186. Rolf, D.: Improved bound for the PPSZ/Schöning-algorithm for 3-SAT. J. Satisfiability, Boolean Modeling and Computation **1**(2), 111–122 (2006)

187. van Rooij, J.M.M., Bodlaender, H.L.: Design by measure and conquer, a faster exact algorithm for dominating set. In: Proceedings of the 25th Annual Symposium on Theoretical Aspects of Computer Science (STACS 2008), LIPIcs, pp. 657–668. Schloss Dagstuhl - Leibniz-Zentrum fuer Informatik, Germany (2008)

188. van Rooij, J.M.M., Bodlaender, H.L.: Exact algorithms for edge domination. In: Proceedings of the 3rd International Workshop on Parameterized and Exact Computation (IWPEC 2009), *Lecture Notes in Comput. Sci.*, vol. 5018, pp. 214–225. Springer (2008)

189. van Rooij, J.M.M., Bodlaender, H.L., Rossmanith, P.: Dynamic programming on tree decompositions using generalised fast subset convolution. In: Proceedings of the 17th Annual European Symposium on Algorithms (ESA 2009), *Lecture Notes in Comput. Sci.*, vol. 5757, pp. 566–577. Springer (2009)

190. van Rooij, J.M.M., Nederlof, J., van Dijk, T.C.: Inclusion/exclusion meets measure and conquer. In: Proceedings of the 17th Annual European Symposium on Algorithms (ESA 2009), *Lecture Notes in Comput. Sci.*, vol. 5757, pp. 554–565. Springer (2009)

191. Rosen, K.: Discrete Mathematics and its applications. McGraw-Hill (1999)

192. Rota, G.C.: On the foundations of combinatorial theory. I. Theory of Möbius functions. Z. Wahrscheinlichkeitstheorie und Verw. Gebiete **2**, 340–368 (1964)

193. Rudin, W.: Real and complex analysis, third edn. McGraw-Hill Book Co., New York (1987)

194. Ryser, H.J.: Combinatorial mathematics. The Carus Mathematical Monographs, No. 14. Published by The Mathematical Association of America (1963)

195. Savitch, W.J.: Relationships between nondeterministic and deterministic tape complexities. J. Comput. System. Sci. **4**, 177–192 (1970)

196. Scheder, D.: Guided search and a faster deterministic algorithm for 3-SAT. In: Proceedings of the 8th Latin American Symposium on Theoretical Informatics (LATIN 2008), *Lecture Notes in Comput. Sci.*, vol. 4957, pp. 60–71. Springer (2008)

197. Schiermeyer, I.: Efficiency in exponential time for domination-type problems. Discrete Appl. Math. **156**(17), 3291–3297 (2008)

198. Schönhage, A., Strassen, V.: Schnelle Multiplikation grosser Zahlen. Computing (Arch. Elektron. Rechnen) **7**, 281–292 (1971)
199. Schöning, U.: A probabilistic algorithm for k-SAT and constraint satisfaction problems. In: Proceedings of the 40th Annual Symposium on Foundations of Computer Science (FOCS 1999), pp. 410–414. IEEE (1999)
200. Schöning, U.: A probabilistic algorithm for k-SAT based on limited local search and restart. Algorithmica **32**(4), 615–623 (2002)
201. Schöning, U.: Algorithmics in exponential time. In: Proceedings of the 22nd International Symposium on Theoretical Aspects of Computer Science (STACS 2005), *Lecture Notes in Comput. Sci.*, vol. 3404, pp. 36–43. Springer (2005)
202. Schroeppel, R., Shamir, A.: A $T = O(2^{n/2})$, $S = O(2^{n/4})$ algorithm for certain NP-complete problems. SIAM J. Comput. **10**(3), 456–464 (1981)
203. Scott, A.D., Sorkin, G.B.: Linear-programming design and analysis of fast algorithms for Max 2-CSP. Discrete Optimization **4**(3-4), 260–287 (2007)
204. Seymour, P.D., Thomas, R.: Graph searching and a min-max theorem for tree-width. J. Combin. Theory Ser. B **58**(1), 22–33 (1993)
205. Sipser, M.: The history and status of the P versus NP question. In: Proceedings of the 24th annual ACM Symposium on Theory of Computing (STOC 1992), pp. 603–618. ACM (1992).
206. Sipser, M.: Introduction to the Theory of Computation. International Thomson Publishing (1996)
207. Stanley, R.P.: Enumerative combinatorics. Vol. 1, *Cambridge Studies in Advanced Mathematics*, vol. 49. Cambridge University Press, Cambridge (1997)
208. Stearns, R.E., Hunt III, H.B.: Power indices and easier hard problems. Math. Systems Theory **23**(4), 209–225 (1990).
209. Stepanov, A.: Exact algorithms for hard counting, listing and decision problems. Ph.D. thesis, University of Bergen (2008)
210. Strassen, V.: Gaussian elimination is not optimal. Numer. Math. **13**, 354–356 (1969)
211. Suchan, K., Villanger, Y.: Computing pathwidth faster than 2^n. In: Proceedings of the 4th International Workshop on Parameterized and Exact Computation (IWPEC 2009), *Lecture Notes in Comput. Sci.*, vol. 5917, pp. 324–335. Springer (2009)
212. Sunil Chandran, L., Grandoni, F.: Refined memorization for vertex cover. Inf. Process. Lett. **93**(3), 125–131 (2005)
213. Tarjan, R.E., Trojanowski, A.E.: Finding a maximum independent set. SIAM J. Comput. **6**(3), 537–546 (1977)
214. Telle, J.A.: Complexity of domination-type problems in graphs. Nordic Journal of Computing **1**, 157–171 (1994)
215. Vassilevska, V., Williams, R.: Finding, minimizing, and counting weighted subgraphs. In: Proceedings of the 41st annual ACM Symposium on Theory of Computing (STOC 2009), pp. 455–464. ACM (2009).
216. Williams, R.: A new algorithm for optimal 2-constraint satisfaction and its implications. Theor. Comp. Sci. **348**(2-3), 357–365 (2005)
217. Williams, R.: Algorithms and resource requirements for fundamental problems. Ph.D. thesis, Carnegie Mellon University (2007)
218. Williams, R.: Applying practice to theory. SIGACT News **39**(4), 37–52 (2008)
219. Williams, R.: Improving exhaustive search implies superpolynomial lower bounds. In: Proceedings of 42th ACM Symposium on Theory of Computing (STOC 2010), pp. 231–240. ACM (2010)
220. Woeginger, G.: Exact algorithms for NP-hard problems: A survey. In: Combinatorial Optimization - Eureka, you shrink!, *Lecture Notes in Comput. Sci.*, vol. 2570, pp. 185–207. Springer (2003)
221. Woeginger, G.J.: Open problems around exact algorithms. Discrete Appl. Math. **156**(3), 397–405 (2008).
222. Yates, F.: The Design and Analysis of Factorial Experiments. Harpenden (1937)

Appendix: Fundamental Notions on Graphs

A *graph* is a pair $G = (V, E)$ of sets such that E is a set of 2-elements subsets of V. The elements of V are the *vertices* and the elements of E are the edges of G. Sometimes the vertex set of a graph G is referred to as $V(G)$ and its edge set as $E(G)$. In this book graphs are always finite, i.e. the sets V and E are finite, and simple, which means that not two elements of E are equal. Unless specified otherwise, we use parameters $n = |V|$ and $m = |E|$. An edge of an undirected graph with endpoints u and v is denoted by $\{u, v\}$; the endpoints u and v are said to be *adjacent*, and one is said to be a *neighbor* of the other. In a directed graph an edge going from vertex u to vertex v is denoted by (u, v). Sometimes we write edge $\{u, v\}$ or (u, v) as uv.

The *complement* of undirected graph $G = (V, E)$ is denoted by $\overline{G}$; its vertex set is V and its edge set is $\overline{E} = \{\{u, v\} : \{u, v\} \notin E, u \neq v\}$. For any non-empty subset $W \subseteq V$, the subgraph of G induced by W is denoted by $G[W]$; its vertex set is W and its edge set consists of all those edges of E with both endpoints in W. For $S \subseteq V$ we often use $G \setminus S$ to denote the graph $G[V \setminus S]$. We also write $G \setminus v$ instead of $G \setminus \{v\}$. The *neighborhood* of a vertex v is $N(v) = \{u \in V : \{u, v\} \in E\}$ and the *closed neighborhood* of v is $N[v] = N(v) \cup \{v\}$. For a vertex set $S \subseteq V$ we denote by $N(S)$ the set $\bigcup_{v \in S} N(v) \setminus S$. We denote by $d(v)$ the *degree* of a vertex v. The minimum degree of a graph G is denoted by $\delta(G)$. The maximum degree of a graph G is denoted by $\Delta(G)$. A graph G is called *r-regular* if all vertices of G have degree r. A 3-regular graph is also called a *cubic* graph.

A *walk* of length k is a non-empty graph $W = (V, E)$ of the form

$$V = \{v_0, v_1, \ldots, v_k\} \quad E = \{v_0 v_1, v_1 v_2, \ldots, v_{k-1} v_k\}.$$

A walk is a *path*, if the v_i are all distinct. If $P = v_0 v_1 \ldots v_k$ is a path, then the graph obtained fro P by adding edge $x_k x_0$ is called a *cycle* of length k. A *Hamiltonian path (cycle)* in a graph G is a path (cycle) passing through all vertices of G. We denote by $d(v, w)$ the *distance* between v and w in the graph G, which is the shortest length of a path between v and w. For any integer $k \geq 1$ and any vertex v of G, we denote by $N^k(v)$ the set of all vertices w satisfying $d(v, w) = k$.

A graph G is connected if for every pair u, v of its vertices there is a path between u and v. A *tree* T is a connected graph without cycles. A *forest* F is a graph without cycle; thus all the connected components of F are trees. A *spanning tree* T of a graph G is a tree such that $V(T) = V(G)$ and $E(T) \subseteq E(G)$.

An *independent set* I of a graph $G = (V, E)$ is a subset of the vertex set V such that the vertices of I are pairwise non-adjacent. The maximum size of an independent set of a graph G is denoted by $\alpha(G)$. A *clique* C of a graph $G = (V, E)$ is a subset of the vertex set V such that the vertices of C are pairwise adjacent. By $\omega(G)$ we denote the maximum clique-size of a graph G. Let us remark that $\alpha(G) = \omega(\overline{G})$. A *dominating set* D of a graph $G = (V, E)$ is a subset of the vertex set V such that every vertex of $V \setminus D$ has a neighbor in D. By $\gamma(G)$ we denote the minimum size of a dominating set of a graph G.

A *coloring* of a graph G assigns a *color* to each vertex of G such that adjacent vertices receive distinct colors. The *chromatic number* of G denoted by $\chi(G)$ is the minimum k such that there is coloring of G using k colors. A domatic partition of a graph G is a partition of the vertex set of G into dominating sets $V_1, V_2, \ldots, V_k$. The *domatic number* of a graph G denotes the minimum number of dominating sets in any domatic partition of G.

A *vertex cover* C of a graph $G = (V, E)$ is a subset of the vertex set V such that C covers the edge set E, i.e. every edge of G has at least one endpoint in C. An *edge cover* C of a graph $G = (V, E)$ is a subset of the edge set E such that C covers the vertex set V, i.e. every vertex of G is endpoint of at least one of the edges in C. A *matching* M of a graph $G = (V, E)$ is a subset of the edge set E matching!perfect such that no two edges of M have a common endpoint. A *perfect matching* is a matching M covering all vertices of the graph, i.e. every vertex is endpoint of an edge in M.

Two graphs $G = (V, E)$ and $H = (W, F)$ are *isomorphic*, denoted by $G \cong H$, if there is a bijection $I : V \to W$ such that for all $u, v \in V$ holds $\{u, v\} \in E \Leftrightarrow \{f(u), f(v)\} \in F$. Such a bijection I is called an *isomorphism*. If $G = H$, it is called an *automorphism*. A mapping $h : V \to W$ is a *homomorphism* from graph $G = (V, E)$ to graph $H = (W, F)$ if for all $u, v \in V$: $\{u, v\} \in E$ implies $\{f(u), f(v)\} \in F$.

For more information on Graph Theory we refer to the textbooks by Bondy and Murty [33], Diestel [65] and Berge [18].

Index